Statistics for the Terrified

Statistics for the Terrified

Gerald Kranzler
University of Oregon

Janet Moursund
University of Oregon

Prentice Hall, Englewood Cliffs, New Jersey 07632

Library of Congress Cataloging-in-Publication Data

KRANZLER, GERALD D.

Statistics for the terrified/Gerald D. Kranzler, Janet P. Moursund.

p. cm.
Includes index.
ISBN 0-13-183831-8
1. Statistics. I. Moursund, Janet. II. Title
QA276.12.K73 1995
519.5—dc20 94-34406
CIP

Acquisition editor: *Heidi Freund*
Editorial/production supervision: *Kim Gueterman*
Cover design: *Helen Woczyk*
Buyer: *Tricia Kenny*

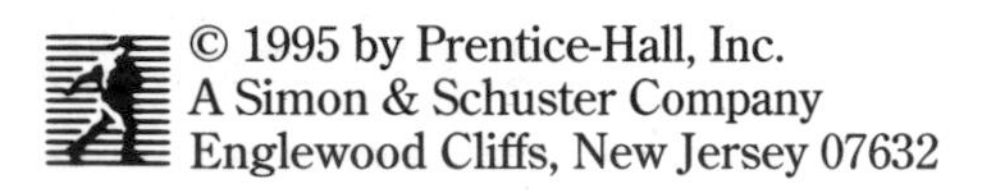

A Simon & Schuster Company
Englewood Cliffs, New Jersey 07632

Printed in the United States of America
10 9 8 7 6 5 4 3 2 1

ISBN 0-13-183831-8

Prentice-Hall International (UK) Limited, *London*
Prentice-Hall of Australia Pty. Limited, *Sydney*
Prentice-Hall Canada Inc, *Toronto*
Prentice-Hall Hispanoamericana, S.A., *Mexico*
Prentice-Hall of India Private Limited, *New Delhi*
Prentice-Hall of Japan, Inc., *Tokyo*
Simon & Schuster Asia Pte. Ltd., *Singapore*
Editora Prentice-Hall do Brasil, Ltda., *Rio de Janeiro*

Statistics for the Terrified has grown over nearly 20 years of working with terrified and not-so-terrified students. In many ways, it is much more their book than ours. It is to these students—the real authors—that this book is dedicated.

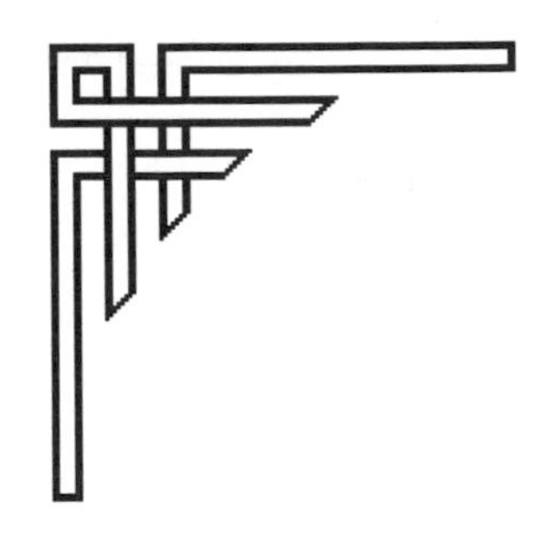

Contents

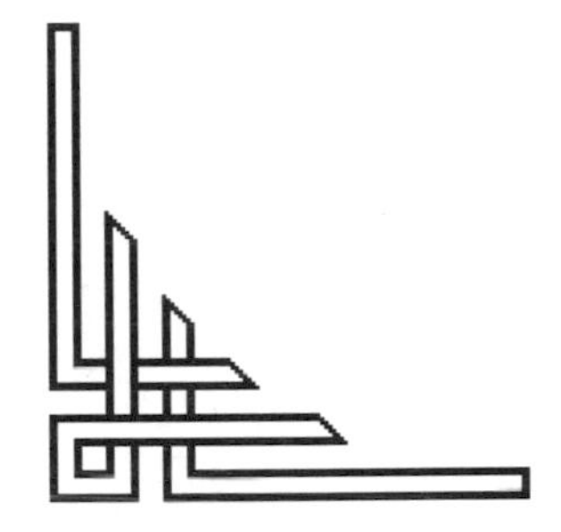

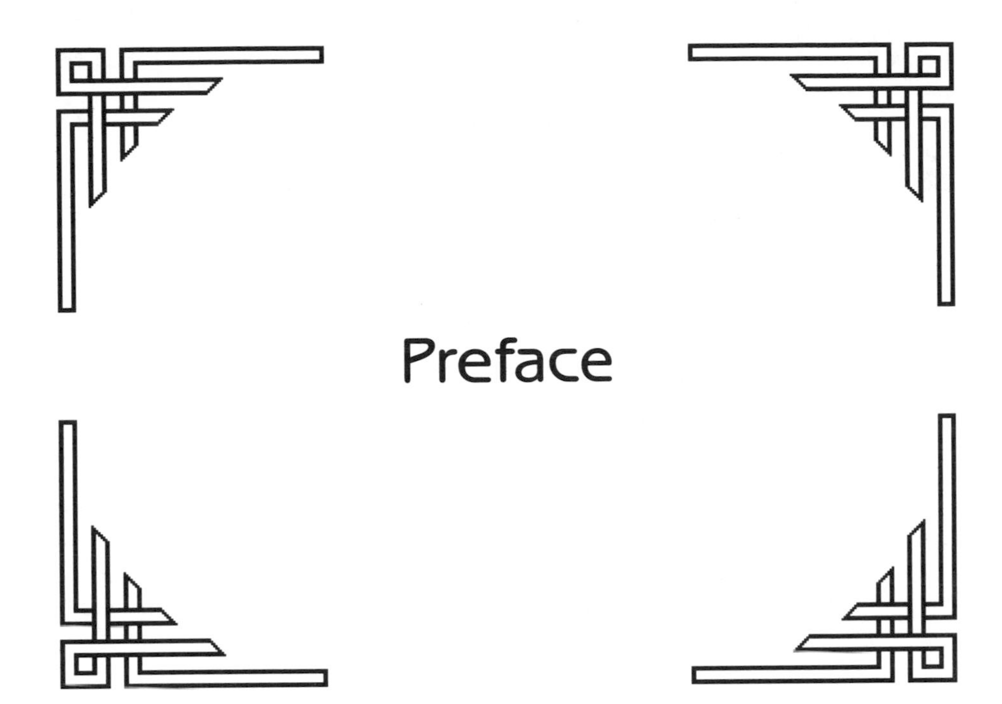

Preface

A book like this, one that has grown over so many years, has made friends with a lot of people. And I want to acknowledge those people, to say "thank you" for their nudges and pats-on-the-back and occasional less friendly but equally useful bits of information.

When I've read other books—especially testbooks—I've tended to skip over all that "front stuff" and I expect most readers do the same thing. I tried to think of some way to make this section witty, gripping, suspenseful—something to make you, the reader, stay with it until the very end. But I couldn't figure out how to do it. So I decided, instead, to simply ask: please, take just a few minutes to read over the list of folks who've helped grow this book. They deserve it.

First on the list, of course, should be all the students. If you read the dedication, you'll know why. I'm not going to list their names but they know who they are, and how much they've contributed, and how much I appreciate it. There have been colleagues, too, over the years, who have supported our efforts and encouraged me to keep going. Amoung them the name that stands our most is that of Suzie Prichard, who calls herself a secretary but is enormously more than that. Without her, this book would never have made it out of the bottom drawer and into the Prentice Hall mail-room.

Somewhere in the editorial halls of good old P-H there was a chain of folks who passed the manuscript on until it got into the hands of the reviewers. I know of Pete Janzow and of Marylin Coco; there may be others. And the reviewers themselves: Clint D. Anderson at Providence College, James L. Dannemiller at the University of Wisconsin—Madison, and Peter J. Rowe at the College of Charleston. Their comments weren't always fun to read, but they were always helpful—even when I didn't follow their suggestions to the letter!

This book's very own editor, Heidi Freund, and its production manager, Kim Gueterman, have been consistently helpful. Roses to them!

And the artist, Katie Tate—she took fuzzy ideas and turned them into wonderful pictures. Katie has been competent, professional, and incredibly patient thoughout the whole process of making this book. She's added immeasurable to its readability, and I am in her debt.

There, are others, of course—there always are—friends and family and even small dog Zak who attacked the mailperson so gallantly when she tried to deliver the galley proofs to my door, but this has to stop somewhere. So finally, thanks to you, Dear Reader, for using my book. Read it, cuss at it, throw it against the wall, but hang in there with me and maybe you will end up (dare I hope?) even liking statistics. Just a bit.

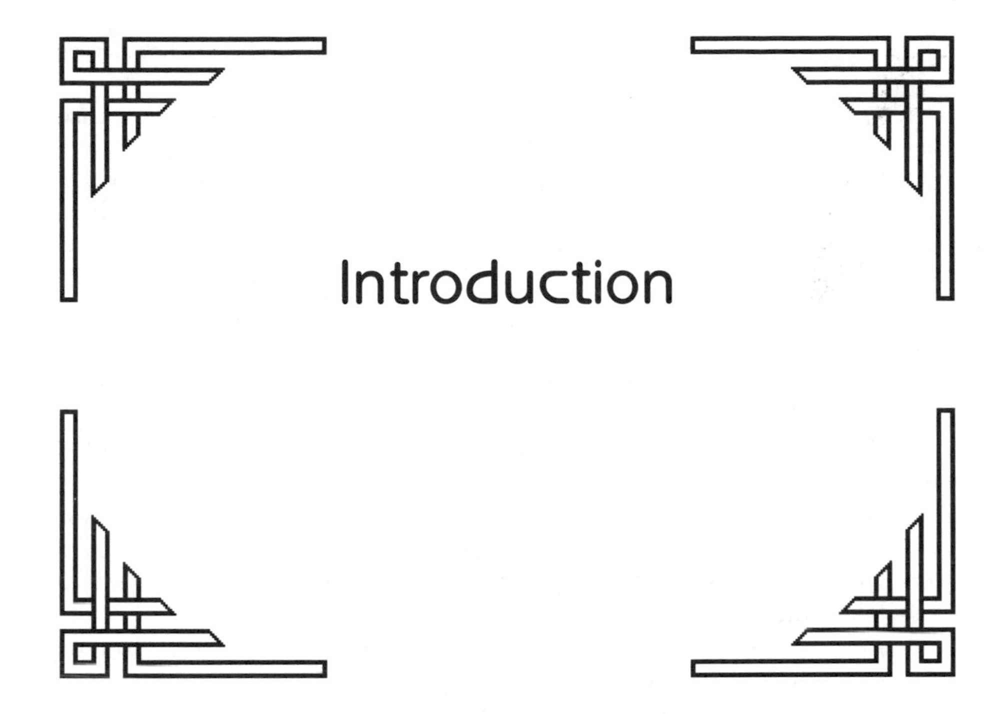

Introduction

"You haven't told me yet," said Lady Nuttal, "what it is your fiancee does for a living."

"He's a statistician," replied Lamia, with an annoying sense of being on the defensive.

Lady Nuttal was obviously taken aback. It had not occurred to her that statisticians entered into normal social relationships. The species, she would have surmised, was perpetuated in some collateral manner, like mules.

"But Aunt Sara, it's a very interesting profession," said Lamia warmly. "I don't doubt it," said her aunt, who obviously doubted it very much. "To express anything important in mere figures is so plainly impossible that there must be endless scope for well-paid advice on how to do it. But don't you think that life with a statistician would be rather, shall we say, humdrum?"

Lamia was silent. She felt reluctant to discuss the surprising depth of emotional possibility which she had discovered below Edward's numerical veneer.

"It's not the figures themselves," she said finally, "it's what you do with them that matters."

K.A.C. Manderville, *The Undoing of Lamia Gurdleneck**

Another statistics book! There are so many statistics books on the market now that it seems strange even to me that there be another one. However,

*Alas, and with apologies to Mr. Manderville, the source of this quotation has been lost in the mists of time. We would dearly like to know its origin, and if any reader can help us out we would be most eternally grateful.

as a teacher of statistics, I have been dissatisfied with available books because they seem aimed at students who whizzed right through college algebra and considered taking math as a major just for the sheer joy of it. Most of my students in Counseling and Education programs are not like that. Many of them would respond with a hearty "true" to many of the following self-test statements. I invite you to test yourself, to see if you too fit the pattern.

1. I have never been very good at math.
2. When my teacher tried to teach me long division in the fourth grade, I seriously considered dropping out of school.
3. When we got to extracting square roots, thoughts of suicide flashed through my mind.
4. Word problems! My head felt like a solid block of wood when I was asked to solve problems like, "If it takes Mr. Jones three hours to mow a lawn and Mr. Smith two hours to mow the same lawn, how long will it take if they mow it together?"
5. Although I never dropped out of school, I became a quantitative dropout soon after my first algebra course.
6. I avoided courses like Chemistry and Physics because they required math.
7. I decided early that there were some careers I could not pursue because I was poor in math.
8. When I take a test that includes math problems, I get so nervous that my mind goes blank and I forget all the material I studied.
9. Sometimes I wonder if I am a little stupid.
10. I feel nervous just thinking about taking a statistics course.

Did you answer "true" to many of these items? If so, this book may be helpful to you. When writing it, I made some negative and some positive assumptions about you:

1. You are studying statistics only because it is a requirement in your major area of study.
2. You are terrified (or at least somewhat anxious) about math and are sure you cannot pass a course in statistics.
3. It has been a long time since you last studied math, and what little you knew then has long since been forgotten.
4. But with a little instruction and a lot of hard work on your part, you can learn statistics. If you can stay calm while baking a cake or balancing your checkbook, there is hope for you.
5. You may never learn to love statistics, but you can change your self-concept. When you finish your statistics course you will be able to say,

truthfully, "I am the kind of person who can learn statistics! I'm not stupid."

In this book I will attempt to help you to achieve two important objectives: (1) how to understand and compute some basic statistics, and (2) how to deal with math anxiety and avoidance responses that interfere with your learning.

Here is some advice that I think you will find useful as we move along: (1) Because the use of statistics requires you to work with numbers, you should consider buying a calculator. Make sure that the calculator has at least one memory and that it can take square roots (almost all calculators can do this). (2) Try to form a support group of fellow statistics students. Exchange telephone numbers and times when you can be reached. Talk about what you are studying and offer to help others (you may learn best by teaching others). When you are stuck with a problem that you can't solve, don't hesitate to ask others for their help. Very likely they have some of the same feelings and difficulties you do. Not everyone gets stuck on the same topics, so even you may be helpful to someone else.

If you are one of the "terrified" for whom this book is intended, there are two appendixes at the end of the book that may be helpful to you. Appendix J, Overcoming Math Anxiety, will give you some general tools and techniques for dealing with the uncomfortable feelings that many students experience when they find themselves dealing with numbers. And Appendix A provides a review of some of the basic math concepts that you may have once known, but that have gotten rusty through disuse. It also gives some sample questions that will allow you to test your ability to use those concepts. I know that reading an appendix before you even get to the first chapter of a book may seem pretty weird (and probably not Politically Correct), but I think these may help you to get off to a running start and will be well worth your time and trouble. Of course, if you don't have problems with math, and already know all the basics, you won't learn anything new; but, even so, "Shucks, I know all this" is a great way to begin a statistics class. Especially if you think it may be terrifying!

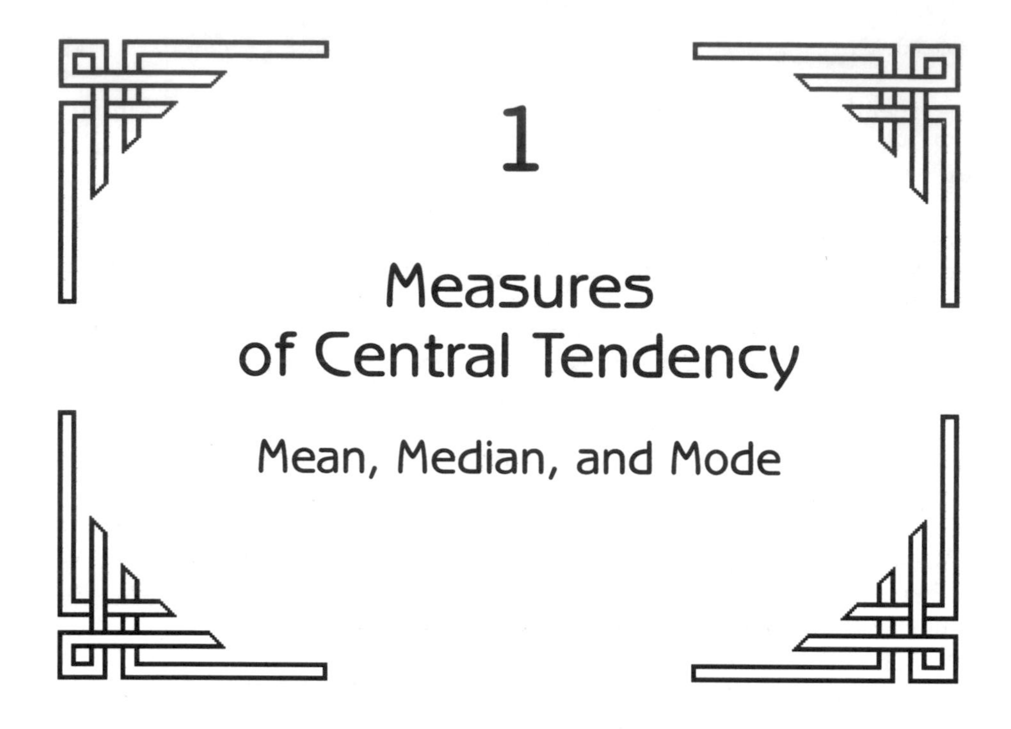

1

Measures of Central Tendency

Mean, Median, and Mode

The field of statistics is concerned with methods of organizing, summarizing, and interpreting data. This section will focus on the "summarizing" part. It will present methods of finding one number that best represents a group of numbers.

We often ask questions such as, "What is the average IQ of this class?" or "How much money does the average Oregon counselor make?" When we ask such questions, sometimes we are not aware that there is more than one "average." In fact, in this section you will study three methods for computing an average: the mean, the median, and the mode.

Before presenting the mean, median, or mode, however, I need to digress a bit and discuss symbolizing data.

SYMBOLIZING DATA

In the next few paragraphs you will be introduced to quantitative methods of summarizing data. If you have read, or even looked at, any statistics books, you must have noticed the use of many symbols unfamiliar to you. Learning to read statistics material is somewhat analogous to learning to read music or a foreign language: impossible at first, difficult for a while, but relatively easy after some effort.

Variables

Those of us who are interested in education or the behavioral sciences are concerned mostly with the characteristics of people, such as intelligence, achievement, interest, or personality. When you study people, one of the first things you will notice is that people vary considerably on almost any human characteristic. Some people are much brighter than others, some learn more than others in the same amount of time, some people's personality characteristics are considerably different from others. In statistics we call such characteristics *variables.*

It is conventional to designate variables by capital letters near the end of the alphabet. For example, the intelligence test scores earned by a group of five students could be designated with the capital letter X, while achievement test scores earned by the same students could be represented by a Y. The first student's intelligence test score would be X_1; the third student's achievement test score would be Y_3. With X and Y as your two variables, you could ask, "What is the average score on the X variable?" or "Is there any relationship between X and Y variables?"

The Summation Sign

You will also be introduced to quite a few Greek letters, especially sigma, designated by Σ. The Greek letter sigma (Σ) directs you to sum (add up) whatever comes after it. If intelligence test scores are represented by the letter X, ΣX directs you to add up all the X scores:

Example

IQ Scores (X)
105
133
113
94
112
$\Sigma X = 557$

Similarly, if achievement test scores are represented by the letter Y, and the test scores are 82, 71, 69, 50, and 22, then $\Sigma Y = 82 + 71 + 69 + 50 + 22 = 294$. You will find more information about working with the summation sign in Appendix A, Basic Math Review.

Parentheses

Strange as it may seem, parentheses are an important mathematical symbol. They serve as a kind of recipe, telling us in what order to do things. In a complicated set of cooking directions, it can be important to know

whether to add the sugar before you beat the egg whites or afterward; in the same way, it's important to know whether to square a set of numbers before or after you add them together.

The parentheses rule is simple: work from the inside out. Carry out whatever operations are inside the innermost set of parentheses, and then whatever is inside the next set, and so on. For example, $(((x(x + y)) - 32)/y)^2$ means to (1) add x and y together; (2) multiply the sum by x; (3) subtract 32; (4) divide by y; and finally (5) square your answer. The parentheses are important because doing the operations in a different order will give you a different result. Try it. Let $x = 2$ and $y = 5$, and see what happens when you change the order of operations!

THE MEAN

The *mean* is the most often used measure of central tendency (*central tendency* is a fancy statistical term that means, roughly, "middleness"). The mean is an old acquaintance of yours: the arithmetic average. You obtain the mean by adding up all the scores and dividing by the number of scores. Remember?

The mean of a variable is represented by a bar over the letter symbolizing it.[1] For example, if the intelligence test scores of a group of students are represented by X, the mean intelligence test score of those students will be represented as $\overline{X}$. If achievement test scores are represented with a Y, the mean achievement test score will be $\overline{Y}$.

The formula for the mean of variable X is

$$\overline{Y} = \frac{\sum Y}{N}$$

This formula says, in words, that the mean of variable X (symbolized as $\overline{X}$) equals the sum of the X scores (ΣX) divided by the number of scores (N).

Similarly, for the scores on the Y variable, the mean of Y is

$$\overline{Y} = \frac{\sum Y}{N}$$

Example

IQ Scores (X)	Achievement Test Scores (Y)
105	14
133	19
113	13
94	9
112	13
$\Sigma X = 557$	$\Sigma Y = 68$
$\overline{X} = 557/5 = 111.4$	$\overline{Y} = 68/5 = 13.6$

Have you noticed how complicated it was to describe the mean in words, compared with that short little formula? Formulas, and mathematical relationships in general, often don't translate into words easily. Mathematicians are trained to think in terms of relationships and formulas and often don't have to translate; we do. That's one reason why social science folks can have problems with statistics: we don't realize that we need to translate, and that the translating takes time. We expect to read and understand a page in a stat book

[1]By convention, $\overline{X}$ and $\overline{Y}$ (pronounced "x-bar" or "y- bar") are used to designate the mean of a *sample,* that is, a finite, countable, measurable set of values. Sometimes we want to refer to the mean of a less definite, often uncountable set of values: the mean IQ score of all the fifth-graders in the United States, for example. A large, inclusive group like this is called a ***population,*** and its mean is symbolized by the Greek letter μ (pronounced "mew," like a kitten). Values having to do with populations are usually symbolized using lowercase Greek letters; for sample values we use the normal English-language alphabet.

To be technically correct, we would have to define a ***population*** as the collection of all the things that fit the population definition and a ***sample*** as some specified number of things selected out of the population. You'll see why that's important when we talk about inferential statistics in Chapter 7. For now, though, just assume that we are working with samples—relatively small, measurable groups.

as quickly as a page in any other sort of book. Not so! Math simply takes longer to read, and we need to remember to slow ourselves down. On the average, you can expect a page of statistics to require as much reading time as three or four pages of nonmathematical information. So don't beat yourself up for being slow—you're supposed to be that way!

THE MEDIAN (Md)

When scores are arranged in order, from highest to lowest (or lowest to highest), the *median* is the middle score. Suppose you administered an IQ test to five persons who scored as follows: 113, 133, 95, 112, 94. To find the median, you would first arrange all scores in numerical order and then find the middle score.

Example

IQ Scores	
133	
113	
112	median = Md = 112
95	
94	

In our example, 112 is the median score because, when scores are arranged in order (highest to lowest), two scores are higher than 112 and two scores are lower than 112.

But suppose you have six scores:

Example

IQ Scores
105
102
101
92
91
80

In this example the number 101 can't be the Md because there are two scores above it and three below. Nor can the number 92 be the Md, for similar reasons. The problem is solved by taking the two middle scores—in our example, 101 and 92—and finding the point halfway between. You do this by adding the two middle scores and dividing by 2: 101 + 92 = 193, divided by 2 = 96.5 = Md. (Did you notice that this is the same as finding the mean of the two middle scores? Good for you!) The Md of our six scores is 96.5. As you can see, now there are three scores that are higher than 96.5 and three that are lower.

Here's another example. Find the median of the following scores: 27, 12, 78, 104, 45, 34. First, arrange the scores from highest to lowest: 104, 78, 45, 34, 27, 12. Then, find the point half way between the two middle scores. The two middle scores are 45 and 34. Half way between is 45 + 34 = 79, divided by 2 = 39.5. Thus, Md = 39.5. Duck soup, right?

THE MODE (Mo)

The *mode* (Mo) is the most frequently occurring score in a set of scores. For example, we are given the following scores:

Example

IQ Scores	
110	
105	
100	
100	Mo = 100
100	
99	
98	

Because the number 100 occurs more frequently than any of the other scores, Mo = 100.

But what about the following set of scores?

Example

IQ Scores
110
105
105
105
100
95
95
95
90

In this example both 105 and 95 occur three times. Here we have what is known as a *bimodal* distribution of scores. If there were more than two modes, it would be a *multimodal* distribution.

SELECTING A MEASURE OF CENTRAL TENDENCY

It has been said that statistics don't lie, but statisticians do. Paraphrased, we could say that statistics can be employed to enhance communication or to deceive.

Consider Ruritania, a country so small that its entire population consists of five persons, a king and four subjects. Their annual incomes are as follows:

	Income in Dollars
King	1,000,000
Subject 1	5,000
Subject 2	4,000
Subject 3	4,000
Subject 4	2,000

The king boasts that Ruritania is a fantastic country with an "average" annual income of $203,000. Before rushing off to become a citizen, you would be wise to find out what measure of central tendency he is using! True, the mean is $203,000, so the king is not lying, but he is not telling a very accurate story either. In this case, either the median or the mode would be more representative values. The point to be made here is that your selection of a measure of central tendency will be determined by your objectives in communication as well as by mathematical considerations.

I will discuss selection of a measure of central tendency again in Chapter 3, in the context of normal and skewed curves. Don't spend any time worrying about that now, though.

PROBLEMS

1. Compute the mean, median, and mode for each of these distributions.

A	B	C	D
3	2	1	2
3	2	3	3
4	2	3	4
6	5	3	4
7	5	5	4
8	7	5	5
10	7	8	7
	8	8	8
	10	8	8
	11	9	
		11	

Answers

1-A. $\overline{X} = 5.86$; Md = 6; Mo = 3
1-B. $\overline{X} = 5.9$; Md = 6; Mo = 2
1-C. $\overline{X} = 5.82$; Md = 5; Mo = 3 and 8
1-D. $\overline{X} = 5$; Md = 4; Mo = 4

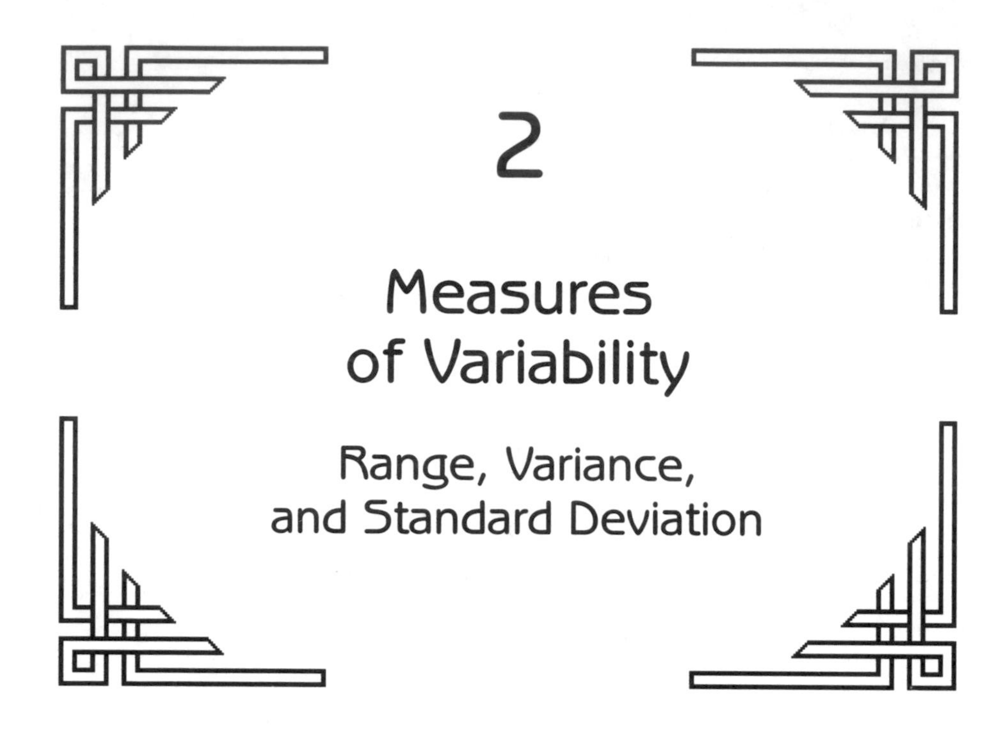

2

Measures of Variability

Range, Variance, and Standard Deviation

In Chapter 1, when we computed measures of central tendency (mean, median, or mode), we wanted one score that would best represent an entire set of scores. Consider the IQs earned by students in each of two classrooms:

Classroom A IQ Scores	Classroom B IQ Scores
160	102
130	101
100	100
70	99
40	98
$\Sigma X = 500$	$\Sigma X = 500$
$X = \frac{500}{5} = 100$	$X = \frac{500}{5} = 100$

Notice that the mean is 100 in both classrooms. But what a difference in variability! (Perhaps you have heard about the man who drowned in a lake with an average depth of 1 foot.) Fortunately for us (?), statisticians have developed several measures of variability that allow us to differentiate between groups of scores like those above.

THE RANGE

The simplest measure of variability is the range. The *range* is the highest score (H) minus the lowest score (L). In classroom A,

$$\text{range} = H - L = 160 - 40 = 120$$

In classroom B,

$$\text{range} = 102 - 98 = 4$$

Because the range is based on only the two most extreme scores, it can be quite misleading as a measure of overall variability. Remember Ruritania, where the king had an annual income of $1,000,000 and the other four people in the sample had incomes of $5000, $4000, $4000, and $2000? The range of this distribution is $998,000, even though all but one of the people in the sample are clustered within $3000 of each other. In this distribution, the range is not as useful a measure as the variance, which is based on all the scores. That's where we go next.

THE VARIANCE (S^2)

The *variance* is the most frequently used measure of variability. The formula for the variance may seem quite intimidating to you at first, but you can handle it if you follow the procedures outlined below. Here's the formula:

$$S^2 = \frac{\sum\left(X - \overline{X}\right)^2}{N}$$

Before learning how to compute the variance, though, let's discuss the concept of deviations about the mean.

Deviations about the Mean

We can tell how far each score deviates from the mean by subtracting the mean from it, using the formula $(X - \overline{X})$. Notice that in the following distribution of scores we have subtracted the mean (5) from each score:

Scores	$X - \overline{X}$
9	$9 - 5 = +4$
7	$7 - 5 = +2$
5	$5 - 5 = 0$
3	$3 - 5 = -2$
1	$1 - 5 = -4$

$\sum X = 25$ (sum) $\sum\left(X - \overline{X}\right) = 0$ (sum of (score − mean))

$\overline{X} = 5$ (mean)

Notice also that when we add up the column headed with $X - \overline{X}$, the sum of that column equals zero. That is, $\Sigma(X - \overline{X}) = 0$. Another way of expressing this relationship is to say that *the sum of the deviations of scores about their mean is zero.* This generalization is always true, except when you make rounding errors. In fact, another definition of the mean is *that score around which the sum of the deviations equals zero.* The sum of the deviations about the mean would not, therefore, make a very good measure of variability because it would be the same for every distribution.

If we square each deviation, however, we get rid of the minus signs. In the following distribution, look carefully at the column headed $(X - \overline{X})^2$. Notice that squaring the deviations gets rid of the negative values.

Scores (X)	$(X-\overline{X})$	$(X-\overline{X})^2$
9	$9-5=+4$	16
7	$7-5=+2$	4
5	$5-5=\ \ 0$	0
3	$3-5=-2$	4
1	$1-5=-4$	16
$\Sigma X=25$	$\Sigma(X-\overline{X})=\ \ 0$	$\Sigma(X-\overline{X})^2=40$

$$\overline{X}=\frac{25}{5}=5$$

Now we have $\Sigma(X - \overline{X})^2 = 40$, the numerator of the formula for the variance. To complete the computation, just divide by N:

$$S^2 = \frac{\sum(X - \overline{X})^2}{N} = \frac{10}{5} = 2$$

Many students complete the computation of their first variance and then ask, "What does it mean?" Perhaps you have a similar question. Before answering it, let's go back to classrooms A and B (from the beginning of the chapter) and find the variance for each classroom.

Classroom A IQ Scores			Classroom B IQ Scores		
Score	$(X-\overline{X})$	$(X-\overline{X})^2$	Score	$(X-\overline{X})$	$(X-\overline{X})^2$
160	60	3600	102	2	4
130	30	900	101	1	1
100	0	0	100	0	0
70	30	900	99	1	1
40	60	3600	98	2	4
500	0	9000	500	0	10

$$\overline{X} = \frac{500}{5} = 100 \qquad \overline{X} = \frac{500}{5} = 100$$

The variance for classroom A is

$$S^2 = \frac{\sum(X - \overline{X})^2}{N} = \frac{9000}{5} = 1800$$

And the variance for classroom B is

$$S^2 = \frac{\sum(X - \overline{X})^2}{N} = \frac{10}{5} = 2$$

Notice that the values for the variances of classrooms A and B indicate that there is quite a bit of difference between the variabilities of the classrooms. That's what the variance is supposed to do—provide a measure of the variability. The more variability in a group, the higher the value of the variance; the more homogeneous the group, the lower the variance.

Another way to understand the variance is to notice that, in order to find it, you need to add up all the squared deviations and divide by N. Sound familiar? Very similar to the definition of the mean, don't you think? So one way to understand the variance is to think of it as an average deviation squared, or maybe a mean squared deviation.

The formula given above for the variance is a definitional formula. It is both accurate and adequate for small sets of numbers in which the mean turns out to be a whole number, but it is inconvenient for general use. For most purposes it will be better for you to use a mathematically equivalent computational formula. It's not only easier to use, but it minimizes round-off error.

The computational formula for the variance is

$$S^2 = \frac{N\sum X^2 - \left(\sum X\right)^2}{N^2}$$

Just take a few deep breaths and relax as much as possible. Calmly analyze what you see. Notice that you know quite a bit already: You know that N is the total number of scores, and ΣX is the sum of all the scores. There are two terms you haven't seen before $(\Sigma X)^2$ and ΣX^2. Let's look at each one in turn.

$(\Sigma X)^2$: This term directs you to find the sum of the X scores and then square the sum.

X Scores	
10	
9	
8	
7	
6	
$\Sigma X = 40$	$\Sigma X^2 = (40)^2 = 1600$

ΣX^2: This term directs you to square each X score and then sum the squares.

Scores X	Scores Squared X^2
10	100
9	81
8	64
7	49
6	36
$\Sigma X = 40$	$\Sigma X^2 = 330$
$(\Sigma X)^2 = 1600$	

Note that ΣX^2 is not the same as $(\Sigma X)^2$. It is very important that you make this distinction!

Here are two rules that may help you to read statistical formulas:

Rule 1. Whenever you see parentheses around a term, as in $(\Sigma X)^2$, do what's indicated inside the parentheses first before doing what's indicated outside the parentheses. In the last example, you would find ΣX first and then square it: $(\Sigma X)^2 = (40)^2 = 1600$.

Rule 2. When there are no parentheses, a symbol and its exponent are treated as a unit. When you see ΣX^2, first square and then add the squared numbers. In the example above, square each number first and then get the sum of the squares:

$$\Sigma X^2 = 330$$

Another example

Scores X	Scores Squared X^2
140	19,600
120	14,400
100	10,000
80	6,400
60	3,600
$\Sigma X = 500$	$\Sigma X^2 = 54{,}000$
	$(\Sigma X)^2 = 250{,}000$

If you have many numbers to work with, the process of finding the square of each number first, then recording it, and then adding them up is tedious. Here's where you can use your calculator's memory. Follow these steps as applied to the scores in the last example:

1. Clear your calculator's memory.
2. Find the square of 140 and enter it into M+. Do the same for each of your X scores without writing any of the squares down on paper.
3. After you have entered the squares of all scores into M+, push your memory recall button (usually MR) and you should get the correct answer, which is 54,000 in this example. This is the value of ΣX^2.

Meanwhile, back at the variance formula, we still have

$$S^2 = \frac{N\sum X^2 - \left(\sum X\right)^2}{N^2}$$

Let's apply the formula to the following example:

Math Anxiety Scores X
11
9
8
7
6

Just follow these steps:

1. Look at the numerator first (it's as easy as a, b, c).
   a. Find $N\Sigma X^2$. In our example, $N = 5$. To find ΣX^2, remember to square each score first; then sum the squares: $\Sigma X^2 = 351$. Therefore, $N\Sigma X^2 = (5)(351) = 1755$.
   b. Find $(\Sigma X)^2$. (Remember, find ΣX first and then square the result.) $(X)^2 = (41)^2 = 1681$.
   c. Find $N\Sigma X^2 - (\Sigma X)^2$. Use what you found in steps a and b. $1755 - 1681 = 74$. Great! Now you have the numerator.

2. Now look at the denominator. This part is really easy because all you have to do is multiply N by itself. In our example, $N = 5$, so $N^2 = 5 \times 5 = 25$.
3. Now divide the numerator by the denominator to find the variance. The numerator was 74 and the denominator was 25, so S^2 equals 74 divided by 25, which equals 2.96. $S^2 = 2.96$.

To summarize:

$$S^2 = \frac{N\sum X^2 - \left(\sum X\right)^2}{N^2} = \frac{5(352) - 1681}{5^2} = \frac{74}{25} = 2.96$$

Now let's go back and compute the variances for classrooms A and B (from the beginning of the chapter).

Variance of Scores in Classroom A

1. Numerator
   a. $N\Sigma X^2 = (5)(5900) = 295{,}000$
   b. $(\Sigma X)^2 = (500)^2 = 250{,}000$
   c. $N\Sigma X^2 - (\Sigma X)^2 = 295{,}000 - 250{,}000 = 45{,}000$
2. Denominator

$$N^2 = 5^2 = 25$$

3. Computation of S^2

$$S^2 = \frac{45{,}000}{25} = 1800$$

That is, the variance of classroom A is 1800. Notice that this is the same value we got by using the definitional formula. Because the mean was a whole number, there was no round-off error with the definitional formula.

Variance of Scores in Classroom B

1. Numerator
   a. $N\Sigma X^2 = (5)(50{,}010) = 250{,}050$
   b. $(\Sigma X)^2 = (500)^2 = 250{,}000$
   c. $N\Sigma X^2 - (\Sigma X)^2 = 250{,}050 - 250{,}000 = 50$
2. Denominator

$$N^2 = 5^2 = 25$$

3. Computation of S^2

$$S^2 = 50/25 = 2$$

Again, notice that this is the same value you found for the variance of classroom B when you used the definitional formula.

If there is a lesson to be learned here, it is that statistical formulas aren't so much *difficult* as they are *compressed.* There's nothing really hard about following the steps. But we are so used to reading things quickly that we tend to look once at a long formula and then give up, without taking the time to break it into chunks that we can understand. A good rule of thumb is that any line of formula should take about as long to read as a page of text.

THE STANDARD DEVIATION (*S*)

When you read educational and psychological research, you will often come across the term *standard deviation,* designated by the letter S. Once you have found the variance S^2 of any set of scores, finding the standard deviation is easy: just take the square root of the variance. If the variance is 25, the standard deviation will be 5; if the variance is 100, the standard deviation will be 10. To find the standard deviation of classrooms A and B, take the square roots of the variances.

Standard Deviation of Classroom A	Standard Deviation of Classroom B
$S = \sqrt{S^2} = \sqrt{1800} = 42.43$	$S = \sqrt{S^2} = \sqrt{2} = 1.41$

Computing the standard deviation is as easy as falling off a log, right? But what do the numbers mean? Again, as was true with the variance, the values of computed standard deviations indicate the relative variability within a group. When you know that the standard deviation of classroom A, for example, is considerably higher than that of classroom B, this could alert you to the likelihood that it might be harder to teach students in classroom A because of the much greater degree of heterogeneity in that class.

We'll find more uses for the standard deviation in later chapters, so you'll learn more about its meaning at that time. In the meanwhile, take time to practice computing some of the statistics you've learned so far. Find the mean ($\overline{X}$), range (R), variance (S^2), and standard deviation (S) for each of the following groups of scores. Carry out all calculations to three decimal places and round them correct to two places. Compare your answers with mine.

IQ Scores	GPAs	Math Anxiety Scores
160	4.0	27
100	3.6	29
80	3.4	27
70	3.0	20
69	2.5	10
$\overline{X} = 95.80$	3.30	22.60
$S = 33.98$	.514	7.00
$S^2 = 1154.56$	.264	49.04
$R = 91.00$	1.50	19.00

PROBLEMS

Find the range, variance, and standard deviation for the following distributions:

1. 1, 2, 3, 4, 5, 6, 7, 8, 9
2. 10, 20, 30, 40, 50, 60, 70, 80, 90
3. –4, –3, –2, –1, 0, 1, 2, 3, 4
4. .1, .2, .3, .4, .5, .6, .7, .8, .9

Answers

1. R = 8; $S^2 = 6.67$; $S = 2.58$
2. R = 80; $S^2 = 666.67$; $S = 25.82$
3. R = 8; $S^2 = 6.67$; $S = 2.58$
4. R = .8; $S^2 = .067$; $S = .26$

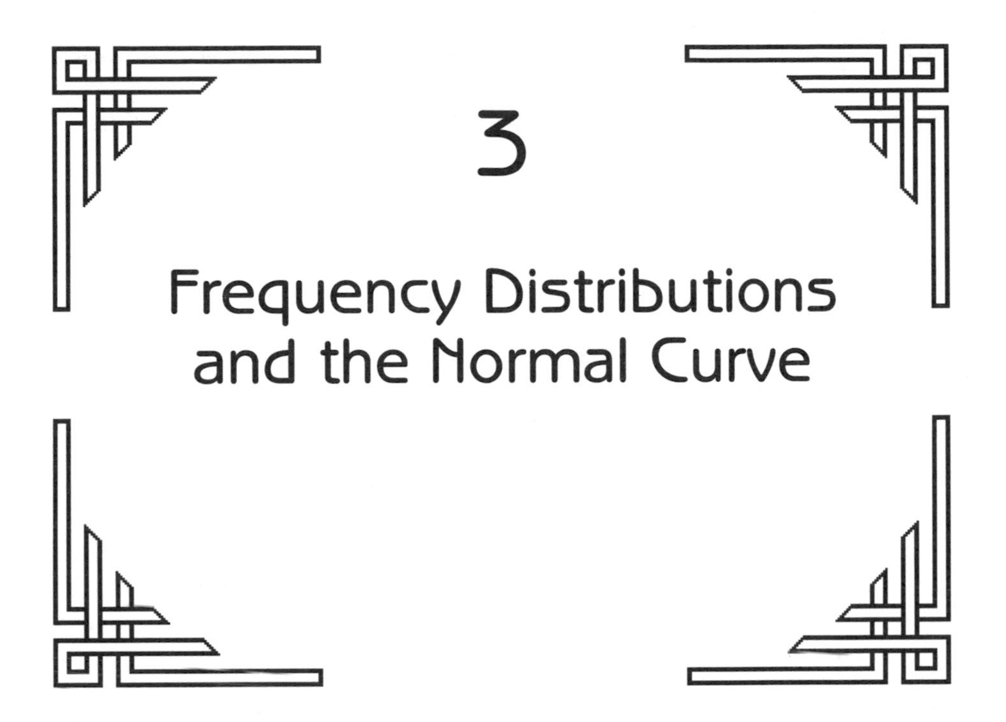

3 Frequency Distributions and the Normal Curve

In the previous chapters you learned some quantitative methods of organizing, summarizing, and interpreting data. You learned how to compute the mean, the median, the mode, the range, the variance, and the standard deviation. You've come a long way! In this chapter you will learn some graphic methods for describing data, and you will put it all together when you focus on the normal curve.

FREQUENCY DISTRIBUTIONS

Imagine that you have administered a math anxiety test to 100 graduate students and that the scores earned by these persons are as listed next (in random order):

Math Anxiety Test Scores Earned by 100 Students

51	50	50	50	51
50	48	49	46	50
45	46	46	47	46
46	46	48	47	46
47	44	49	47	48
49	48	48	49	45
48	46	46	48	48
44	45	44	46	49

Math Anxiety Test Scores Earned by 100 Students (*Continued*)

47	49	43	47	46
47	48	43	48	46
48	46	48	46	47
47	47	47	49	49
46	47	47	44	45
45	48	48	48	47
47	49	47	45	48
49	47	45	47	44
48	47	47	46	47
46	47	46	45	47
45	45	47	48	48
46	48	45	46	47

The simplest way to organize and summarize data such as these is to construct a simple frequency distribution. You can do so by following these steps:

1. Figure out how many different score values there are and write them down in order. By looking over all the scores in our example, you can see that the highest score earned was 51, while the lowest score was 43. List them all from highest to lowest like this:

Math Anxiety Scores
51
50
49
48
47
46
45
44
43

2. Go through the list of scores, score by score, and make a tally each time a score occurs. At the end of this process, your data should look like this:

Math Anxiety Scores	Tally
51	//
50	~~////~~
49	~~////~~ ~~////~~
48	~~////~~ ~~////~~ ~~////~~ ~~////~~
47	~~////~~ ~~////~~ ~~////~~ ~~////~~ ~~////~~
46	~~////~~ ~~////~~ ~~////~~ ~~////~~
45	~~////~~ ~~////~~ /
44	~~////~~
43	//

3. Count up your tallies to find the frequency (f) with which each score was earned. Your complete frequency distribution would look like this:

Math Anxiety Scores	Tally	Frequency (f)
51	//	2
50	~~////~~	5
49	~~////~~ ~~////~~	10
48	~~////~~ ~~////~~ ~~////~~ ~~////~~	20
47	~~////~~ ~~////~~ ~~////~~ ~~////~~ ~~////~~	25
46	~~////~~ ~~////~~ ~~////~~ ~~////~~	20
45	~~////~~ ~~////~~ /	11
44	~~////~~	5
43	//	2

You can see from "eyeball analysis" of your completed frequency distribution that you now have a better picture of how the 100 subjects scored on your math anxiety test than you had when scores were just listed in random order. You can see that most of the scores tend to be bunched up around 47 and that relatively fewer people earned scores at or near the extremes of the distribution.

Data such as these are sometimes presented in graphic form. There are many kinds of graphs, but the most important graph to understand in the study of statistics is the frequency polygon. That's because the frequency polygon forms the basis for understanding the normal curve.

THE FREQUENCY POLYGON

Graphs have a horizontal axis (known as the *x*-axis) and a vertical axis (known as the *y*-axis). It is conventional in statistics to place scores along the *x*-axis and frequencies on the *y*-axis. Let's see how our frequency distribution of math anxiety scores would look in graph form:

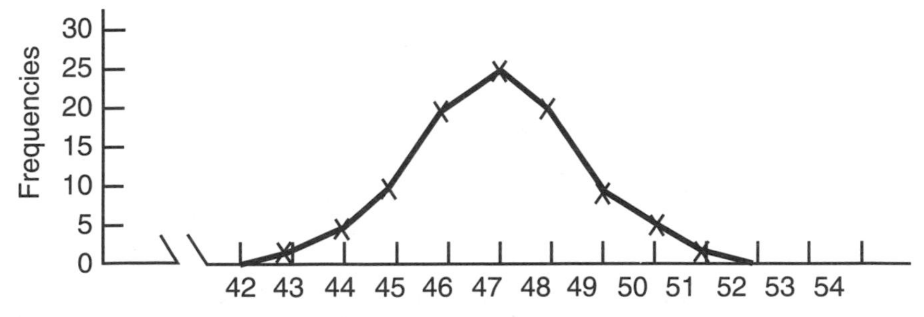

Just in case you haven't worked much with graphs, I'll go through this one slowly.

First, look at the X's. Specifically, look at the X directly above the score of 44. Notice that the X is directly across from the frequency of 5, which indicates that 5 persons earned a score of 44. Similarly, the graph indicates that 20 persons earned a score of 46.[1] Got it? Good!

Second, look at the lines connecting the X's. When constructing a frequency polygon, connect all the X's with a line (you haven't had this much fun since you were a kid). Since the scores 42 and 52 each had zero frequency, you can complete the graph by bringing your line down to the base line at these points, indicating that neither of them had anybody scoring there. Now you have constructed a *frequency polygon*—a many-sided figure, describing the shape of a distribution.

Occasionally, a researcher may want to display data in cumulative form. Instead of building a graph that shows the number of scores occurring at each possible score, a cumulative frequency polygon shows the number of scores occurring at *or below* each point. Here's the *cumulative* frequency polygon for the math anxiety score data:

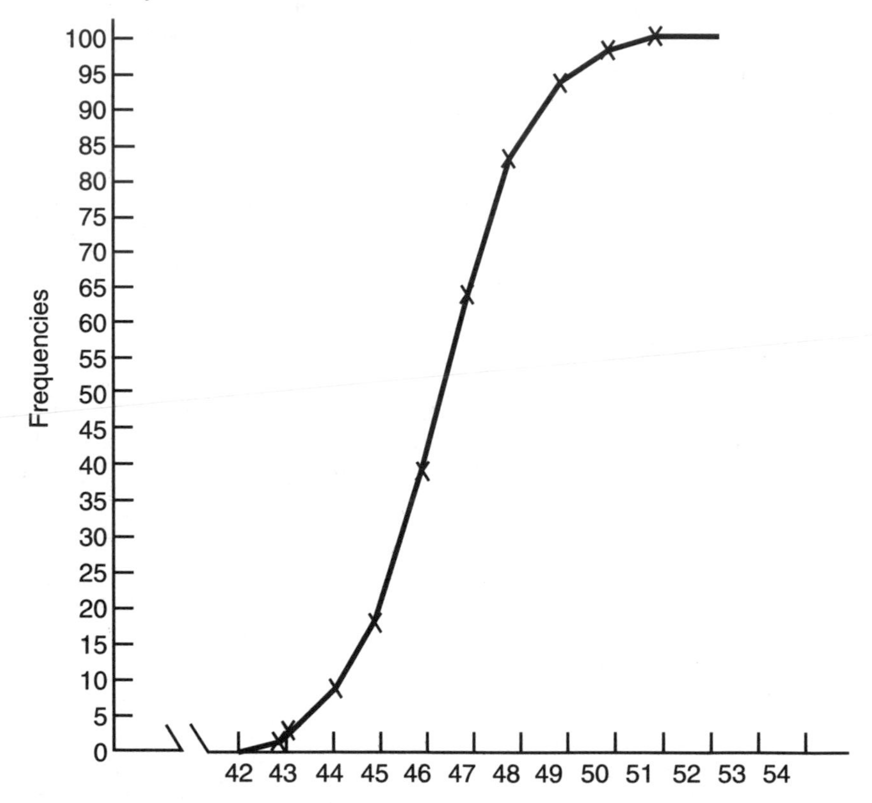

[1]To be perfectly accurate, I should say that 5 people earned a score somewhere between 43.5 (which rounds off to 44) and 44.4 (which also rounds off to 44). Similarly, 20 persons earned a score between 45.5 and 46.4. This way of looking at a range of scores, rather than a single point, isn't very important when we're talking about the anxiety test scores, but when we talk about other kinds of measures (like measurement in inches, which we'll be doing in a minute), it can be very important indeed.

By reading across from any number on the horizontal axis, we can see how many students scored at or below that point. For example, 38 students had anxiety test scores at or below 46.

A cumulative frequency polygon is particularly useful in illustrating learning curves, when a researcher might be interested in knowing how many trials it took for a subject to reach a certain level of performance. Imagine that the graph here was obtained by counting the number of rounds a beginning dart thrower used during a series of practice sessions. The vertical axis is still "frequencies," but now it represents the number of practice rounds; the horizontal numbers represent the thrower's score on any given round. The first two rounds netted him a score of zero—he was figuring out how to hold and throw the darts. He suddenly got the range of the target, and on his third round he scored 43. It took him four more rounds to get a score of 44, a total of seven rounds to that point. By the time he reached a score of 48, he had thrown 83 rounds, and so on.

This cumulative graph is typical of a learning curve: the learning rate is relatively slow at first, picks up speed in the middle, and levels out at the end of the set of trials, producing a flattened S shape.

THE NORMAL CURVE

Around 1870 Quetelet, a Belgian mathematician, and Galton, an English scientist, made a discovery about individual differences that impressed them greatly. Their method was to select a characteristic, such as weight or acuteness of vision, obtain measurements on large numbers of individuals, and then arrange the results in frequency distributions. They found the same pattern of results over and over again, for all sorts of different measurements. The figure on the next page is an example that depicts the results of measuring chest size of over 5000 soldiers.

The rectangles in this graph are called bars (it's a bar graph, or *histogram*), and the bars represent the number of folks who fell into each respective range. About 50 soldiers had chest sizes between 33.0 and 33.999 inches. If we were to put a mark at the top of each bar and drew straight lines between the marks, we'd have a frequency polygon of the sort we drew earlier. The curve that's drawn over the bars doesn't follow that polygon shape exactly, however; it's what we'd get if we measured thousands and thousands more soldiers and plotted the bar graph or frequency polygon for all of them, using narrower measurement intervals—maybe tenths or even hundredths of an inch, instead of whole inches.

The symmetrical, bell-shaped curve that results from plotting human characteristics on frequency polygons closely resembles a curve, familiar to

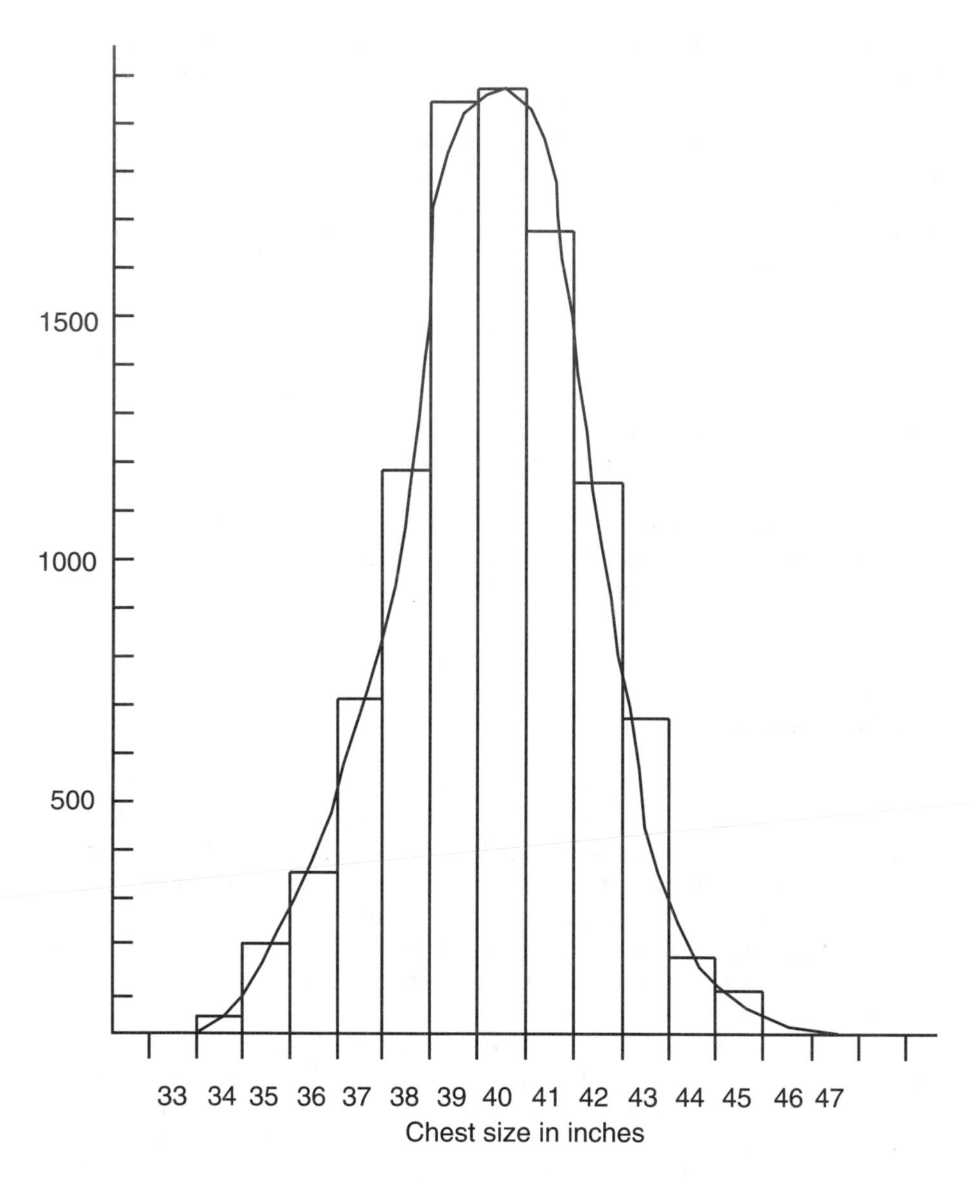

Chest sizes of 5738 soldiers. (Source: Francis Galton, *Natural Inheritance.* London: Macmillan and Co., 1889.)

mathematicians, known as the normal probability curve. The normal curve is bell-shaped and perfectly symmetrical and has a certain degree of "peakedness." Not all frequency polygons have this shape, however. Let's digress for a moment, and talk about skewed curves.

SKEWED CURVES

Skewed curves are not symmetrical. If a very easy arithmetic test were administered to a group of graduate students, for example, chances are that most students would earn high scores and only a few would earn low scores. The scores would tend to "bunch up" at the upper end of the graph. When scores cluster near the upper end of a frequency polygon, the graph is said to be *negatively skewed.* Here's an example of a negatively skewed curve:

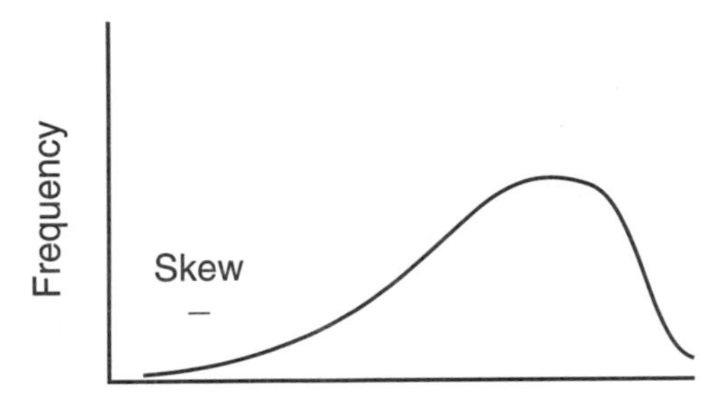

Negatively skewed distribution

On the other hand, if the test were too difficult for the class, most people would get low scores. When graphed on a frequency polygon, these scores would be said to be "*positively skewed.*"

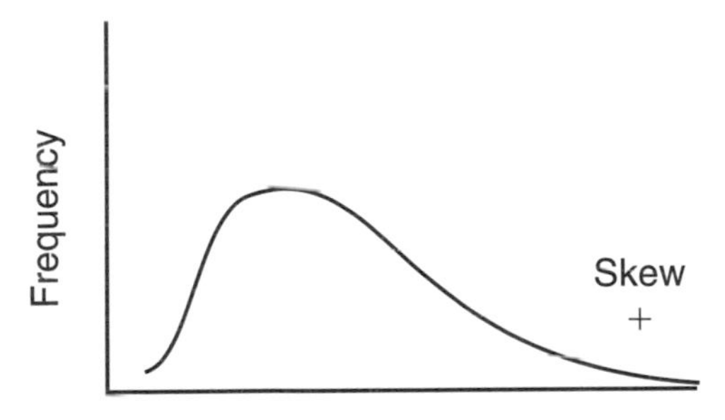

Positively skewed scores

You may recall that, in our discussion of measures of central tendency in Chapter 1, I promised to talk to you later about the choice of mean, median, or mode and the skewness of a distribution. When a distribution is skewed, the mean is most strongly affected. A few scores far out in the tail of a distribution will "pull" the mean in that direction. The median is somewhat "pulled" in the direction of the tail, and the mode is not "pulled" at all. To see what I mean, look at the three distributions on the next page.

The first distribution (*X*) is perfectly symmetrical. Its mode, mean, and median are equal. In unimodal symmetrical distributions,

$$\overline{X} = \text{Md} = \text{Mode}. \qquad \textit{Always.}$$

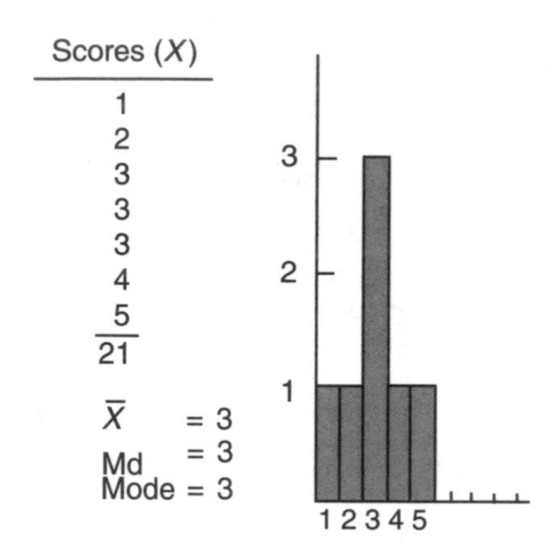
Scores (X)
1
2
3
3
3
4
5
21
$\overline{X}$ = 3
Md = 3
Mode = 3
3
2
1
1 2 3 4 5

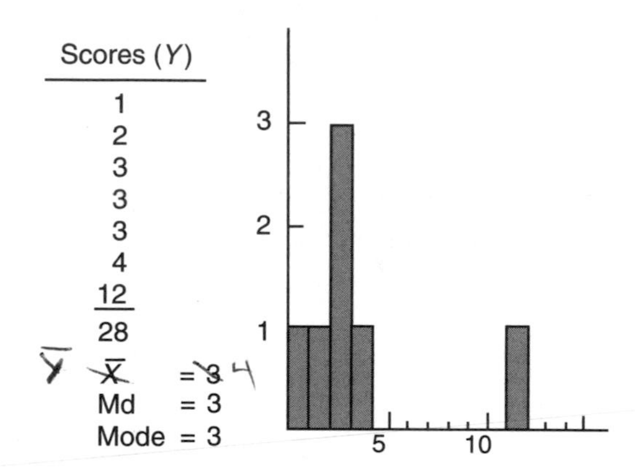
Scores (Y)
1
2
3
3
3
4
12
28
$\overline{X}$ = 3
Md = 3
Mode = 3
3
2
1
5
10

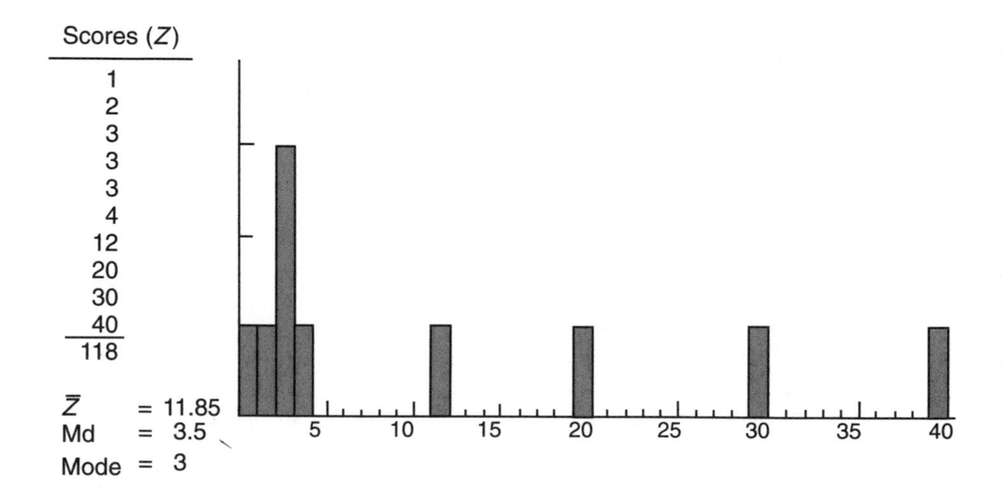
Scores (Z)
1
2
3
3
3
4
12
20
30
40
118
$\overline{Z}$ = 11.85
Md = 3.5
Mode = 3
5
10
15
20
25
30
35
40

Distribution *Y* has been skewed by pulling the largest score, 5, out to 12. This shifted the mean from 3 to 4, but it didn't change either the mode or the median.

Finally, in distribution *Z* we have not only shifted the 5 out to 12, but we have further skewed the distribution by adding scores of 20, 30, and 40. The mean is again most sensitive to these changes. And this time the Md shifts too, just a bit, from 3 to 3.5. But the mode remains unchanged.

Now, after this fascinating discussion, let's get back to our major concern: the normal distribution. Data derived from measurement of human char-

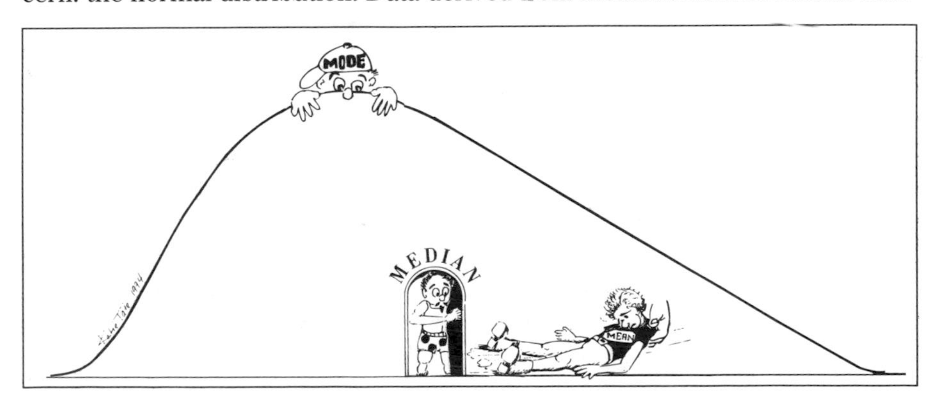

acteristics will never fit the normal probability curve exactly, because the normal probability curve is a hypothetical distribution—it doesn't exist in the real world. But graphs based on frequency distributions of many human characteristics, including psychological test results, often *resemble* the normal curve, at least to some extent, as can be seen in the next graph.

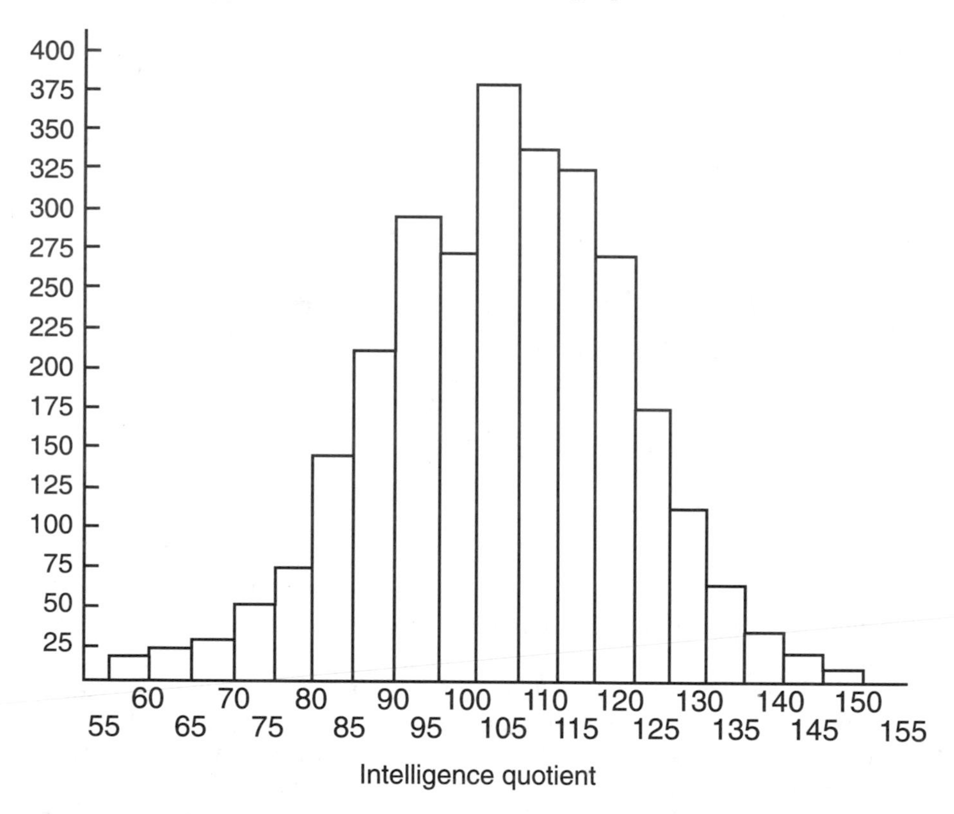

Distribution of IQs on Form L of the Stanford–Binet test for ages 2–18. (Adapted from Quinn McNemar, *The Revision of the Stanford–Binet Scale.* Boston: Houghton Mifflin, 1942.)

Since many actual distributions approximate the normal curve so closely, we can use what mathematicians know about it to help us interpret test results and other data. Notice that, in the graph of IQ scores, large numbers of scores cluster near the middle of the distribution, and there are relatively small numbers of scores near the extremes: most of the scores cluster around 100, while the percent of persons who score lower than 70 or higher than 130 is quite small. When a set of scores is distributed approximately like the normal curve, mathematicians can provide us with a great deal of information about those scores, especially about proportions of scores in different areas of the curve.

PROPORTIONS OF SCORES UNDER THE NORMAL CURVE

In the next few paragraphs I will try to show you how the mean, median, mode, standard deviation, and the normal probability curve all are related to each other. If you are still confused when you finish this section, at least you'll be confused on a higher level.

Consider the following example:

Test Scores (X)	Frequency
110	1
105	2
100	3
95	2
90	1

If you were to calculate the mean, median, and mode from the data in this example, you would find that $X = 100$, Md = 100, and Mo = 100 (go ahead and do it, just for practice). The three measures of central tendency always coincide in any group of scores that is normally distributed.

Recall that the median is the middle score, the score that divides a group of scores exactly in half. For any distribution, you know that 50% of the scores are below the median and 50% above. In a normal distribution the median equals the mean, so you know that 50% of the scores also are higher and lower than the mean. Thus, if you know that the mean of a group of IQ scores is 100, and if you know that the distribution is normal, then you know that 50% of the persons who took the test scored higher than 100 and 50% lower.

Now let's see how the standard deviation fits in. Suppose again that you had administered an IQ test to a very large sample, that the scores earned by that sample were distributed like the normal probability curve, and that the mean equaled 100 ($\overline{X} = 100$) and the standard deviation equaled 15 ($S = 15$). Mathematicians can show that, for scores distributed normally, exactly 34.13% of the scores lie between the mean and one standard deviation away from the mean. (You don't need to know why it works out that way; just take it on faith.) In our example, therefore, 34.13% of the persons' scores would be between 100 and 115 (115 is one standard deviation above the mean: $\overline{X} + S = 100 + 15 = 115$). We know that the normal curve is symmetrical, so we know that 34.13% of the scores will also be between 100 and 85 ($\overline{X} - S = 100 - 15 = 85$). Thus, if we administered our IQ test to 100 persons, approximately 34 would have scores between 100 and 115, and about 68 would have scores between 85 and 115 (34.13% + 34.13%). Most students find it helpful (necessary?) to see a picture of how all of this works; use the following graph to check it out:

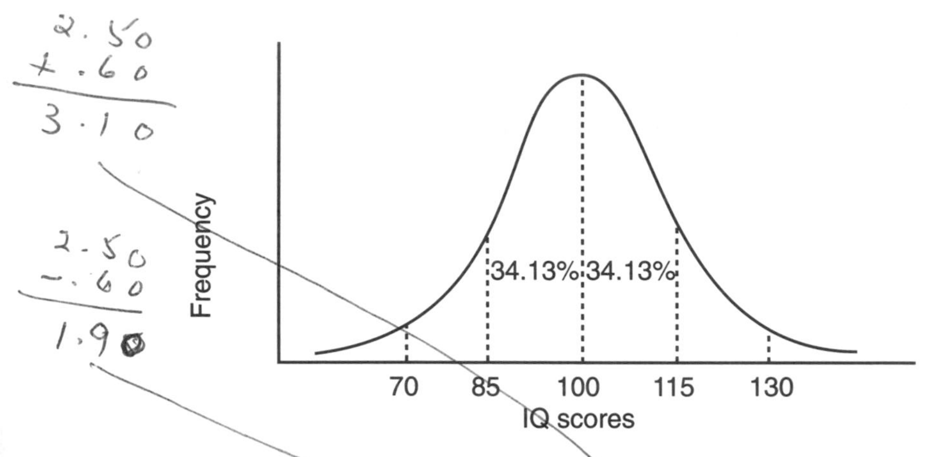

To take another example, suppose that grade-point averages (GPA) were calculated for 1000 students and that $\overline{X} = 2.5$ and $S = .60$. If our sample of GPAs was distributed normally, you would know that 500 students had GPAs higher and 500 had GPAs lower than 2.5. You would also know that approximately 683 of them (34.13% + 34.13% = 68.26% of a thousand, which is approximately 683) had GPAs somewhere between 1.90 and 3.10 (plus and minus one standard deviation). Here's what the graph would look like:

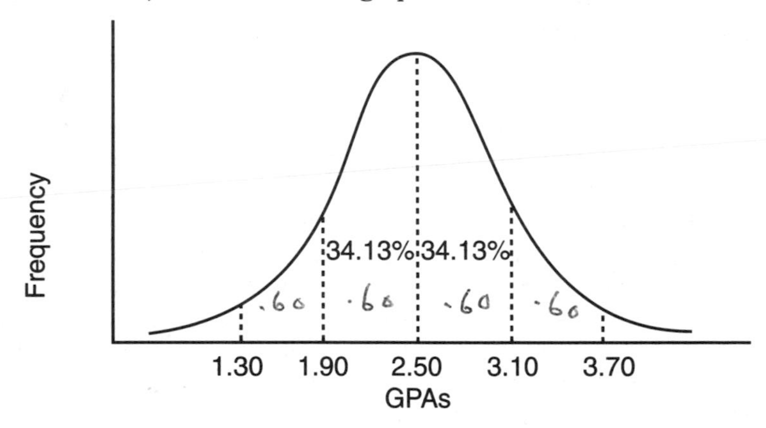

Mathematicians can tell us the proportion of cases between any two points under the normal curve. The figure on the next page presents information about some selected points.

The numbers on the base line of the figure represent standard deviations (S), where –1 represents one standard deviation below the mean, +2 represents two standard deviations above the mean, and so on.

Given the information in the graph, you can answer some interesting questions. For example, suppose again that you administered an IQ test to 100 people, that the scores were distributed normally, and that $X = 100$ and $S = 15$.

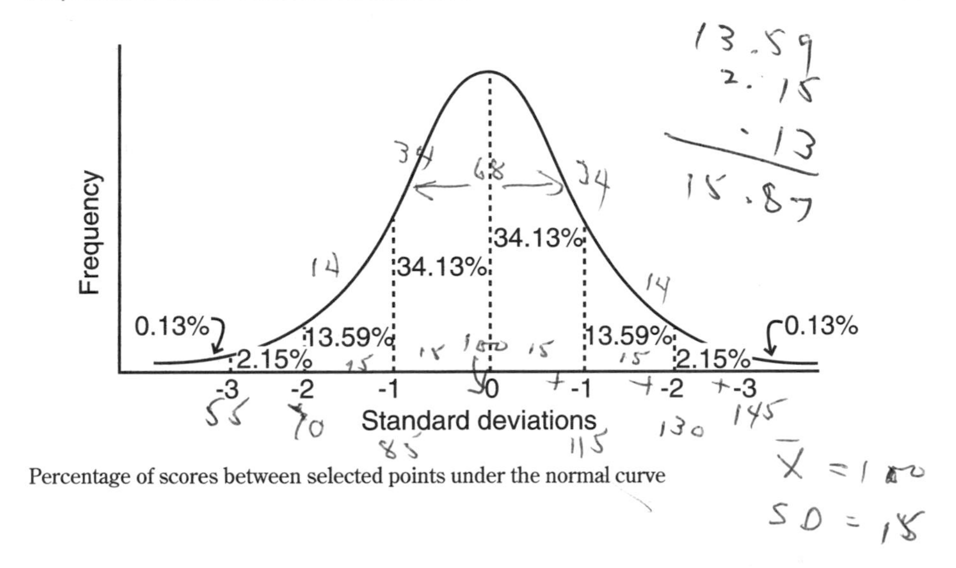

Percentage of scores between selected points under the normal curve

You know from the discussion above that 50 of the 100 people scored lower than 100 and 50 higher. And you know that approximately 68 scored between 85 and 115. The graph also tells you that:

1. 13.59% of the scores are between –1 *S* and –2 *S*. In our example, therefore, around 14 people scored between 70 (two standard deviations below the mean) and 85 (one standard deviation below the mean). Of course, 13.59% of the cases also are between +1 *S* (115) and +2 *S* (130).
2. 95.44% of the cases fall between –2 *S* and +2 *S* (13.59% + 34.14% + 34.13% + 13.59%), so we know that approximately 95 out of 100 people scored between 70 (–2 *S*) and 130 (+2 *S*).
3. 99.74% of the cases fall between –3 *S* and +3 *S*. Almost all persons in our example scored between 55 (+3 *S*) and 145 (+3 *S*).
4. 84.13% had IQs lower than 115. How do we know this? 50% of the persons scored below 100 (the mean and median, remember?), and another 34.13% scored between 100 and 115. Adding the 50% and the 34.13%, you find that 84.13% of the scores were below 115.

The following problems will test your understanding of the relationships among the mean, median, mode, standard deviation, and the normal probability curve. Remember that these relationships hold only if the data you are working with are distributed normally. When you have a skewed distribution or one that is more or less peaked than normal, what you have learned in this chapter does not hold true.

PROBLEMS

Suppose that a test of math anxiety was given to a large group of persons, the scores were normally distributed, that $X = 50$ and $S = 10$. Approximately what percent of persons earned scores:

1. below 50?
2. above 60?
3. below 30?
4. above 80?
5. between 40 and 60?
6. between 30 and 70?
7. between 60 and 70?
8. below 70?
9. below 80?

Answers

1. 50% **2.** 15.87% **3.** 2.28% **4.** .13% **5.** 68.26%
6. 95.44% **7.** 13.59% **8.** 97.72% **9.** 99.87%

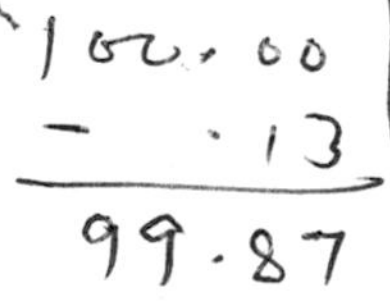

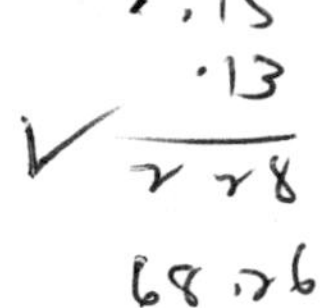

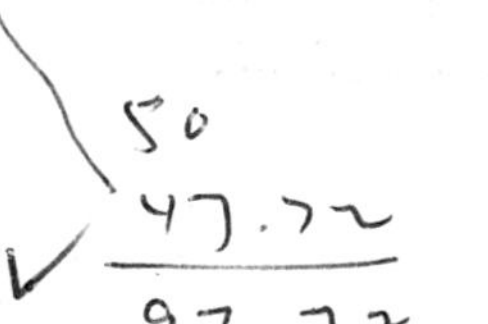

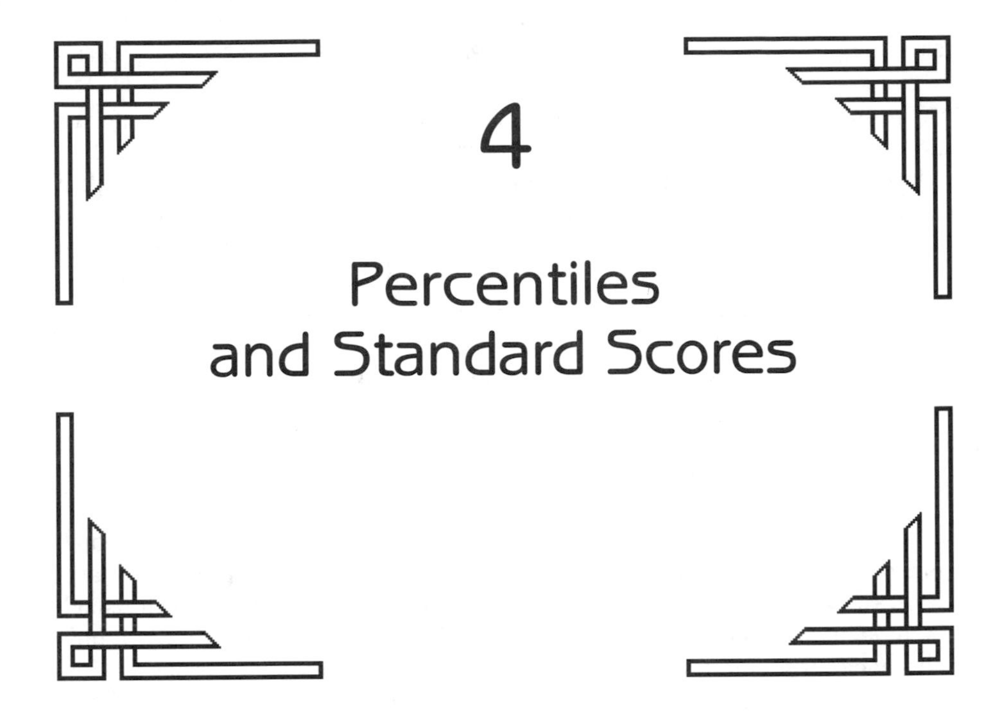

4

Percentiles and Standard Scores

If you work as a teacher or a member of other helping professions, you frequently will be required to interpret material in student or client folders. Material in the folders typically will include results of psychological testing, including written comments, and, more likely, various types of test scores. This chapter will introduce you to two common types of scores—percentiles and standard scores—as well as their major variations. Some of the material will appear complicated at first, but it's just a logical extension of what you've learned so far. You may not even find it particularly difficult!

Before discussing percentiles and standard scores, I want to point out some of the disadvantages of three types of scores with which you may be familiar from your school days: the raw score, the percentage correct score, and rank in class.

> Boy: Ma! I got 98 on my math test today!
> Mother: That's very good, son. You must be very happy!
> Boy: There were 200 points on the test.
> Mother: Oh! I'm sorry. I guess you didn't do too well.
> Boy: I got the second highest score in class.
> Mother: Very good!
> Boy: There are only two of us in the class.

As you can see, the boy's raw score didn't communicate much information. But neither did his percentage correct score, because it didn't tell us whether the test was extremely difficult or very easy. Nor was his rank in class very helpful unless we knew how large the class was, and even when we found that out, we didn't know a whole lot because the class was small and our knowledge of the one person with whom he was being compared is nonexistent. When interpreting his test score, we would like to know, at a minimum, something about the persons with whom he is being compared (the *norm group*) and how he did in comparison with that group.

Man to woman: How's your husband doing?

Woman: Relative to what?

The norm group of a test usually will include large samples of people. For an intelligence test, for example, the publisher will attempt to get a large representative sample of people in general. In their sample they will usually include appropriate proportions of persons in the various age, gender, ethnic, and socioeconomic groups. Test manuals often contain detailed descriptions of normative samples and the way in which they were obtained; a good textbook on tests and measurement can also give you that kind of information.

PERCENTILES

Percentiles are one of the most frequent types of measures used to report the results of standardized tests, and for good reason: they are the easiest kind of score to understand. An individual whose score was at the 75th percentile of a group scored as high or higher than 75% of the persons in the norm group. Someone whose score was at the 50th percentile scored as high or higher than 50% of the persons in the norm group. And someone whose score was at the 37th percentile scored as high or higher than 37% of the persons in the norm group. And so on.

The *percentile rank* of a score in a distribution is the percentage of the whole distribution falling at or below that score. A *percentile* is a measure that tells you how many people (or rats, or rabbits, or whatever the distribution is made up of) scored at or below that point.

Couldn't be much simpler, right?

STANDARD SCORES

On many published psychological tests, raw scores are converted to what are called standard scores. The basic standard score, known as the *Z* score, is defined mathematically by

$$Z = \frac{X - \overline{X}}{S}$$

where X is an individual's raw score
$\overline{X}$ is the mean raw score of the group with which the individual is being compared (usually a norm group of some kind)
S is the standard deviation of that group

Suppose you administered a test to a large number of persons and computed the mean and standard deviation of the raw scores with the following results:

$$\overline{X} = 42$$

$$S = 3$$

Suppose also that four of the individuals tested had these scores:

Person	Score
Jim	45
Sue	48
George	39
Jane	36

What would be the Z score equivalent of each of these raw scores? Let's find Jim's Z score first:

$$Z_{\text{Jim}} = \frac{45 - 42}{3} = +1$$

Notice that (1) we substituted Jim's raw score ($X = 45$) into the formula, and (2) we used the ***group*** mean ($\overline{X} = 42$) and the ***group*** standard deviation ($S = 3$) to find Jim's Z score.

Now for George's Z score:

$$Z_{\text{George}} = \frac{39 - 42}{3} = -1$$

Your turn—you figure out Sue's and Jane's Z scores. Did you get $Z_{\text{Sue}} = +2$, and $Z_{\text{Jane}} = -2$? You did? Very good!

Here is some more practice. Suppose you administered a test to 16 persons who earned the following scores: 20, 19, 19, 18, 18, 18, 17, 17, 17, 17, 16, 16, 16, 15, 15, 14. Start by finding the mean and the standard deviation; I'll give you the answers, but check yourself to see if you can get the same ones:

$$N = 16$$

$$\Sigma X = 272$$

$$(\Sigma X)^2 = 73{,}984$$

$$\Sigma X^2 = 4664$$

$$\overline{X} = 17$$

$$S = 1.58$$

Fred was one of the two persons who scored 19, so for Fred $X = 19$. To find Fred's Z score,

$$Z_{\text{Fred}} = \frac{19-17}{1.58} = +1.27$$

Sarah's raw score (X) was 14. Her Z score is—what? You figure it out.[1]

Notice that the sign in front of the Z score tells you whether the individual's score was above (+) or below (–) the mean.

Another example. Suppose a test was given to a large number of persons with the following results: $X = 47$, $S = 5$. Check your computation of Z scores.

Individual's Raw Score (X)	Z Score
57	+2.0
55	+1.6
52	+1.0
50	+ .6
47	0
45	– .4
42	–1.0
40	–1.4
37	–2.0

What does a Z score of –1.4 tell us? First, the minus sign tells us that the score was below the mean. Second, the number 1.4 tells us that the score was 1.4 standard deviations below the mean.

What about a Z score of +2.0? The plus indicates that the score is above the mean, and the 2.0, again, tells us that the score is 2 standard deviations above the mean.

Notice that whenever a person's raw score is equal to the mean, his or her Z score equals zero. Take the person whose raw score (X) was 47, for example. That person's Z score = 0. If the score had been 1 standard deviation above the mean, the Z score would have been +1.0; if it had been 1 standard deviation below the mean, the Z score would have been –1.0.

To summarize, the Z score tells you if the raw score was above the mean (the Z score is positive) or if the raw score was below the mean (the Z score is negative), and it tells you how many standard deviations the raw score was above or below the mean.

Many people are uncomfortable with negative numbers; others don't like using decimals. Some don't like negatives *or* decimals! Since Z scores often involve both, these folks would rather not have to deal with them. Our

[1]That's right, –1.90

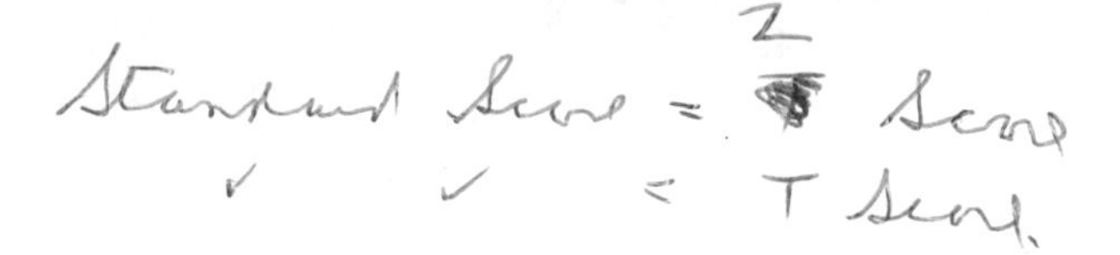

mathematical friends have developed a way to transform Z's into other measures that are always positive and (almost) always use only whole numbers. The most common of these is the T score, which we'll discuss next.

T Scores

A number of published personality inventories report test results in a manner similar to that employed by the California Psychological Inventory (CPI). An example of a CPI profile is shown in Figure 4-1. Along the top of the profile you can see a number of scales designated with letters naming the psychological characteristics measured by the CPI, such as Sc (self-control) and To (tolerance). Under each scale name is a column of numbers. These are raw scores. More interesting to us are the numbers on the left- and right-hand edges. These are the standard score equivalents of the raw scores.

The standard score utilized by the CPI is something called a T score. The formula for T scores is

$$T = 10(Z) + 50$$

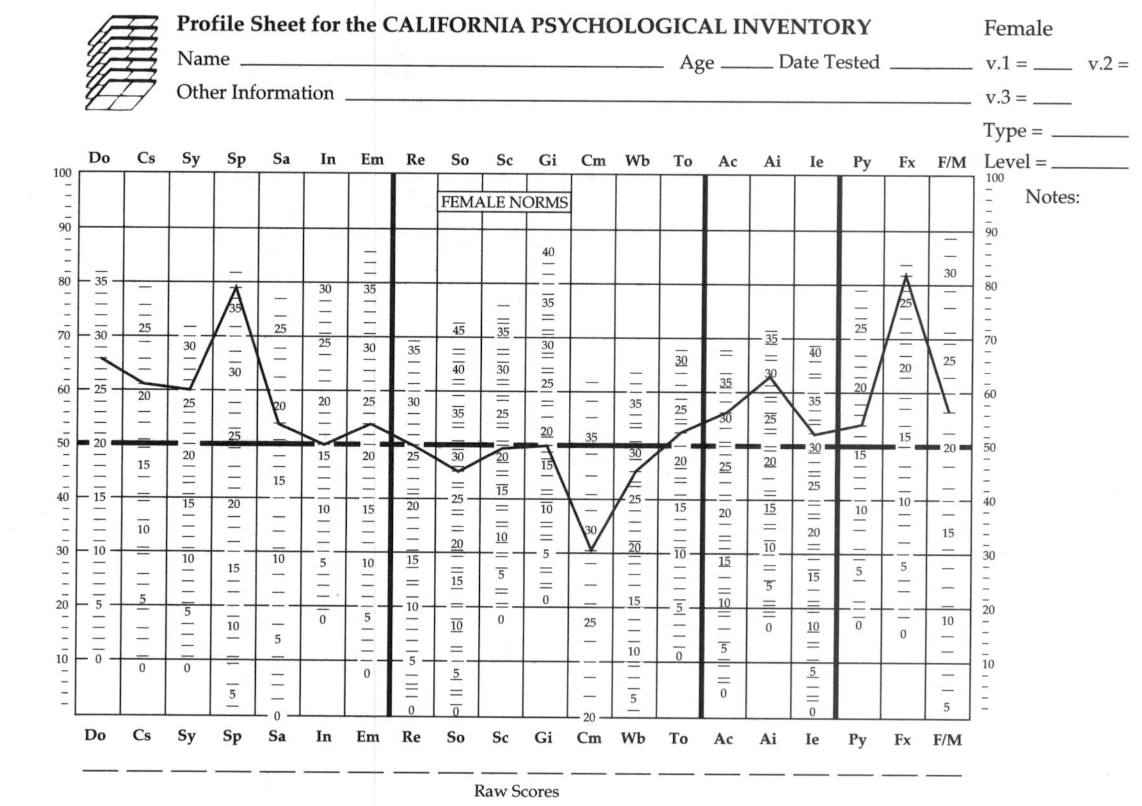

Profile Sheet for the CALIFORNIA PSYCHOLOGICAL INVENTORY
Female
Name
Age
Date Tested
v.1 =
v.2 =
Other Information
v.3 =
Type =
Level =
Notes:
Do Cs Sy Sp Sa In Em Re So Sc Gi Cm Wb To Ac Ai Ie Py Fx F/M
FEMALE NORMS
Raw Scores

Figure 4-1

The Z in this formula is the Z you learned to compute before. Remember that the formula for computing Z scores is

$$Z = \frac{X - \overline{X}}{S}$$

To convert raw scores to T scores, you must:

1. Calculate the $\overline{X}$ and S of the raw scores
2. Find the Z score equivalent of each raw score.
3. Convert the Z scores to T scores by means of the formula $T = 10(Z) + 50$.

Imagine that you administered a test to a large sample of adults and computed the mean and standard deviation of the raw scores. Suppose that $X = 40$ and $S = 3$, and that one of the persons tested, Jim, had a raw score of 46. What was his T score?

1. Step 1 has been done for you; we have found that $\overline{X} = 40$ and $S = 3$.
2. Jim's Z score is

$$Z_{\text{Jim}} = \frac{46 - 40}{3.00} = +2.00$$

3. Jim's T score is $T = (10) \times (+2) + 50 = 20 + 50 = 70$.

Here are the Z and T score equivalents of seven more persons. See if you can get the same answers as I did.

Person	Raw Score	Z Score	T Score
Frank	49	+3.0	80
Sam	40	0.0	50
Sharon	47	+2.3	73
George	42	+0.7	57
Herman	39	–0.3	47
Fred	37	–1.0	40
Kate	32	–2.7	23

Notice that if a person's raw score is right at the mean of the group (see Sam's score) then his or her T score is 50; if the raw score is one standard deviation below the mean (see Fred's score), then his or her T score is 40, and so on.

Look at the sample CPI profile again. Notice that the score on the Sc scale has a T score equivalent of almost exactly 50. This T score lets you know that, relative to the norm group, this person's score was at or near the mean. On the other hand, look at the Sp (social presence) score. The T score is 80, so we know that it is 3 standard deviations above the mean. Very few people in the norm group score that high—the graph in Chapter 3 indicates that well under 1% are up there—so we can conclude that this person had an extremely elevated Sp score.

Deviation IQs

If you took your undergraduate psychology courses some years ago, you were exposed to this IQ formula:

$$IQ = \frac{MA}{CA}(100)$$

In this IQ formula, MA stood for "mental age" and CA for "chronological age." Most modern intelligence tests don't use this formula any more. Instead, most of them use what is known as the ***deviation IQ,*** another form of standard score. Both of the best known individual intelligence tests, the Wechsler scales (WAIS and WISC) and the Stanford–Binet, use deviation IQs, but they're slightly different from one another:

$$\text{Wechsler IQ} = 15(Z) + 100$$

$$\text{Stanford–Binet IQ} = 16(Z) + 100$$

The same raw scores could be used to compute either type of deviation IQ. For example, suppose an intelligence test had been given to a large representative sample of adults, with $\overline{X} = 27$ and $S = 6$. Here are the raw scores (number correct on the intelligence test) earned by six of the persons, their Z scores, and what their equivalent IQs would be if a Wechsler or a Stanford–Binet deviation IQ were computed. (It would be a good idea for you to figure out a few of these on your own, just to prove that you can do it.)

	Raw Score	*Z* Score	Wechsler IQ	Stanford–Binet IQ
Rob	45	+3.00	145	148
Jack	36	+1.50	123	124
Sarah	27	.00	100	100
Fred	20	−1.17	82	81
Abbey	17	−1.67	75	73
Don	8	−3.17	52	49

As you can see, if a person's raw score is exactly at the mean of the raw scores (see Sarah's score), then his or her IQ is 100 for either test. If a score is above the mean, however (see Bob's score), then the Stanford–Binet will be somewhat higher than the Wechsler. And if it's below the mean (see Fred's score), the Wechsler IQ will be higher than the Stanford–Binet.

No matter what test is used or what the raw data look like, using a deviation formula like this guarantees that the average score will be 100 and that about two-thirds of the scores will be between about 85 and about 115. Just like we expect IQs to distribute themselves.

CONVERTING STANDARD SCORES TO PERCENTILES

I hope you have realized by now that the choice of test score to be used by a test publisher is somewhat arbitrary. Some types of score are relatively easy for anyone to understand, while others can be really understood only by those who are sophisticated statistically (like you). My own preference of test-score type is the percentile. Percentile rank tells us exactly where a person stands relative to their norm group, without any need for further translation. A percentile rank of 50 means that the score is exactly in the middle of the norm group—at the median. A percentile rank of 30 means that the score is at the point where 30% of the remaining scores are below it and 70% above it. A percentile rank of 95 means that only 5% of the norm group scores were higher. Don't you just love it when you can score at the 99th percentile on an achievement test?

Figure 4-2 will help you to see the relationship between standard scores (*T*'s, *Z*'s, IQ) and percentiles and the relationship of all these to the normal curve. With a ruler or other straightedge to guide you, you can use the figure to make a rough conversion from one kind of score to another. In a moment, we're going to talk about how to do these conversions mathematically. For now, though, using the chart will help you get a better sense of how it all fits together.

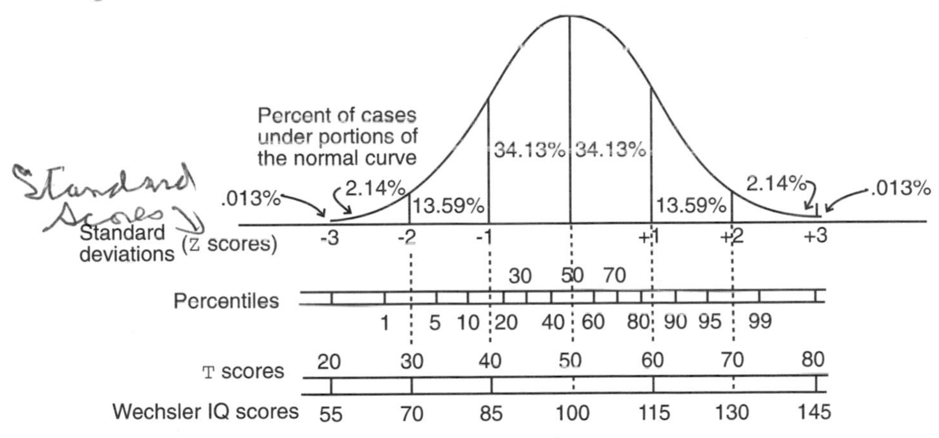

Figure 4-2 Relationships among areas of the normal curve, standard deviations, percentiles, and *Z* and *T* scores.

To use Figure 4-2 to move among different kinds of scores, you first need to convert a raw score to a *Z* score. Having done that, you can easily go to any of the other scores. For example, a score 1 standard deviation above the mean equals

a *Z* score of +1

a *T* score of 60

a deviation IQ of 115 (Wechsler)

a percentile rank of about 84 (84.13)

Similarly, a raw score 2 standard deviations below the mean equals

a Z score of –2

a T score of 30

a deviation IQ of 70 (Wechsler)

a percentile rank of approximately 2 (2.27)

Chances are that it will be helpful at some time in your work to translate standard scores into approximate percentiles. For example, if you know that a person's MMPI Depression score is 70, and if you know that MMPI scores are T scores, then you also know that he or she scored higher than more than 98% of the norm group on that scale. Similarly, if he or she earned an IQ of 85, you will know that the score is at approximately the 16th percentile.

What is the percentile equivalent of someone whose Z score is 1.97? You can answer that question by looking down on Figure 4-2 from the Z score scale to the percentile score and making a rough approximation (a Z score of 1.97 is about the 98th percentile). A more precise answer can be obtained by consulting Appendix B, concerned with proportions of area under the standard normal curve.

Don't panic; I'll tell you how to use that appendix. First, though, remember what you already know: if scores are distributed normally, the mean equals the median. Therefore, the mean is equal to the 50th percentile. Recall also (you can check it on Figure 4-2) that a raw score equal to the mean has a Z score equal to zero. Putting these facts together, you can see that $Z = 0$ = 50th percentile.

You may also recall from Chapter 3 that a score 1 standard deviation above the mean is higher than 84.13% of the norm group, which is another way of saying that a Z score of +1 is at the 84.13th percentile.

Now let's see how to use Appendix B. Look at the following example. Suppose you gave a test to a large group of people, scored their tests, and computed the mean and standard deviation of the raw scores. Assume that the scores were distributed normally and that $\overline{X} = 45$ and $S = 10$. Jim had a raw score of 58. What is his percentile rank?

Procedure for Finding Percentile Ranks Given Raw Scores

First, convert the raw score to a Z score by using

$$Z = \frac{X - \overline{X}}{S}$$

In our example, $X = 58$ (Jim's raw score), $\overline{X} = 45$ (given) and $S = 10$ (given). Therefore,

$$Z_{\text{Jim}} = \frac{58-45}{10} = +1.30$$

At this point, I always draw a picture of a normal (approximately) curve, mark in 1 and 2 standard deviations above and below the mean, and put a check mark where I think the score I'm working with ought to go:

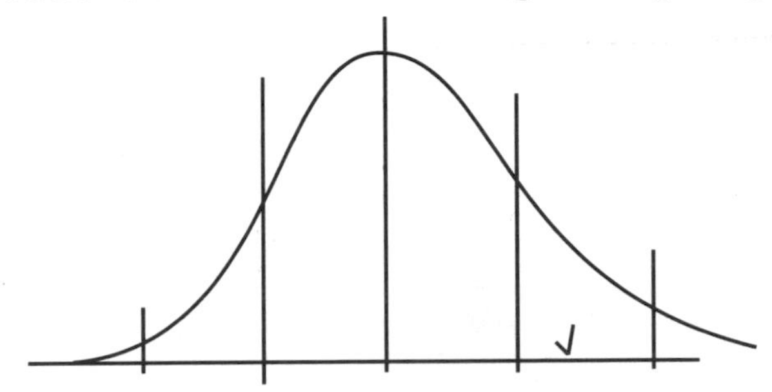

As you can see, it doesn't have to be a very perfect figure. But it gives me an idea of what the answer is going to be. In this case, I know from my picture that the percentile is more than 84 and less than 98[1]; it's probably somewhere in the late 80s. Knowing that helps me to avoid dumb mistakes like reading from the wrong column of the table. If this step helps you, do it. If not, don't.

Now go down the Z column of Appendix B until you come to the Z score you just obtained ($Z = +1.3$), and record the first number to the right (in our example, .4032)

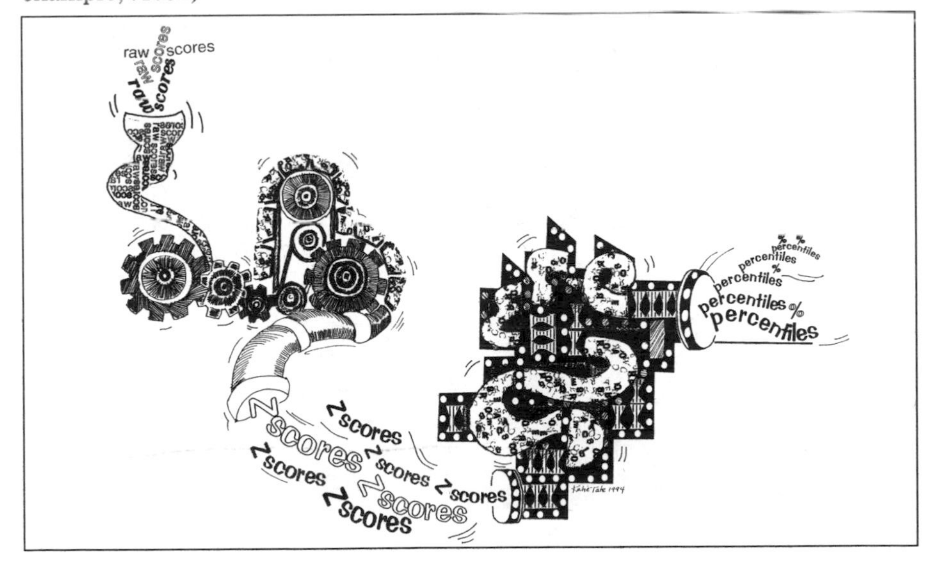

[1]If you don't know where these numbers came from, go back to Figure 4-2.

The obtained number indicates that between the mean and a Z score of +1.3 you will find a proportion of .4032, or 40.32% of the cases (to convert proportions to percents, move the decimal two places to the right). You know that 50% of the cases in a normal curve fall below the mean, so a Z score of +1.3 is as high or higher than .50 + 40.32% = .9032, or 90.32% of the cases. A raw score of 58, therefore, corresponds to a percentile rank of 90.32. That's pretty close to my guess of "late 80s"!

Here's a more accurate picture of what we just did.

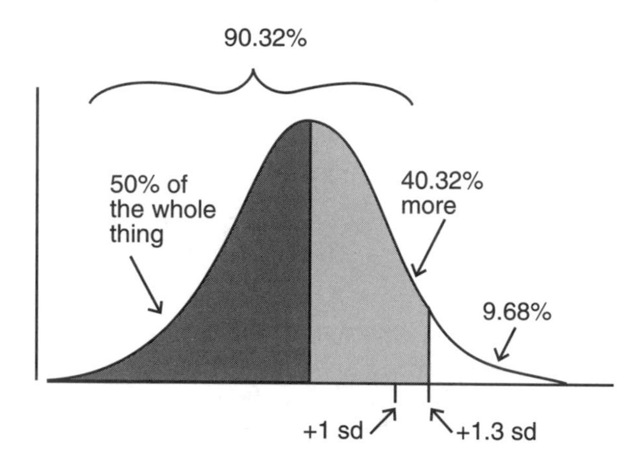

Relationship between a raw score of 58 and the corresponding Z score of +1.30 and the percentile rank of 90.

Another example: In Appendix B, locate a Z score of +0.89. The table indicates that 31.33% of the cases fall between the mean ($Z = 0$) and ($Z = +0.89$), and that 18.67% of the cases fall *above* +0.89. As in the previous problem, the percentile equivalent of $Z = +0.89$ is 50% + 31.33% = 81.33%, a percentile rank of approximately 81.

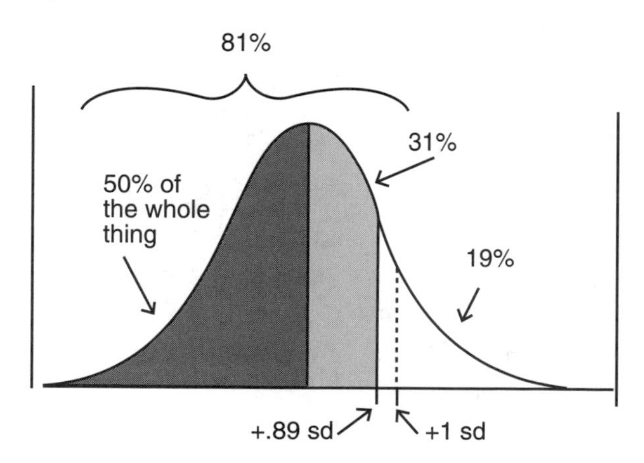

Relationship between a Z score of +0.89 and the corresponding percentile rank of 81.

Remember that the standard normal curve is symmetrical. Thus, even though Appendix B shows areas above the mean, the areas below the mean are identical. For example, the percent of cases between the mean ($Z = 0$) and $Z = -.74$ is about 27%. What is the corresponding percentile rank? Appendix B indicates that beyond $Z = .74$ there are 23% of the cases (third column). Therefore, the percentile rank is 23.

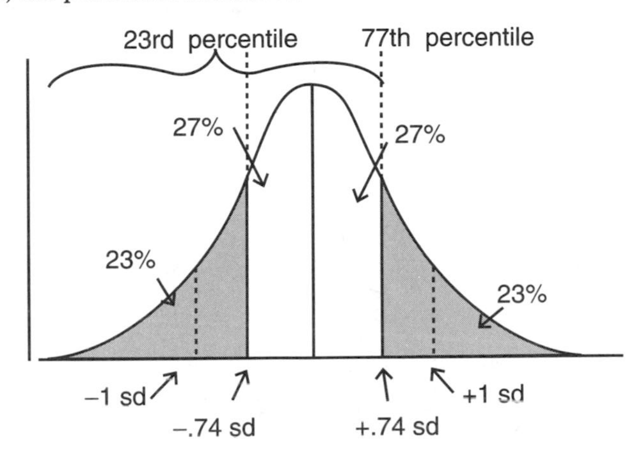

Relationship between a Z score of −.74 and the corresponding percentile rank of 23.

What percent of cases fall between $Z = +1$ and $Z = -1$? Appendix B indicates that 34.13% fall between the mean and $Z = +1$. Again, since the standard normal curve is symmetrical, there are also 34.13% between the mean and $Z = -1$. Therefore, between $Z = -1$ and $Z = +1$ there will be 34.13% + 34.13% = 68.26% of the cases.

Verify for yourself that 95% of the cases fall between $Z = \pm 1.96$ (that is, between $Z = +1$ and $Z = -1$) and that 99% of the cases lie between $Z = \pm 2.58$.

WARNING

Remember that all the material in this chapter assumes that the scores you are interpreting are distributed normally (or nearly normally). If the raw scores are distributed in a skewed fashion, then conversions from standard scores to percentiles will not be accurate. (However, conversions from raw scores to standard scores can always be done. If you feel really ambitious, find a friend and explain why this is so.)

PROBLEMS

1. The following is a hypothetical set of anxiety scale scores from a population of students being seen at a college counseling center:

5	16	47
6	18	50
9	25	50
10	46	50
12	47	78

What are the Z score and the T score of the person who scored 6? 18? Of the two people who scored 47? Of the person who scored 78? Why is it not a good idea to use Appendix B to compute percentiles for these scores?

2. SAT scores are normally distributed, with a mean of 500 and sd of 100. Find the Z, T, and percentile equivalents of the following scores: 500, 510, 450, 460, 650, 660.

Answers

1. In a distribution with $\overline{X} = 31.27$ and $S = 21.7$,

6 (raw score)	= −1.16 (Z score)	= 39 (T score)	
18 (raw score)	= −.6 (Z score)	= 44 (T score)	
47 (raw score)	= .7 (Z score)	= 57 (T score)	
78 (raw score)	= 2.15 (Z score)	= 71 (T score)	

Since the scores are not normally distributed, the table of values for a normal curve can't be used to convert these scores to percentiles.

2.

500 (raw score) =	0 (Z score) =	50 (T score) =	50th percentile
510 (raw score) =	.1 (Z score) =	51 (T score) =	54th percentile
450 (raw score) =	−.5 (Z score) =	45 (T score) =	31st percentile
460 (raw score) =	−.4 (Z score) =	46 (T score) =	35th percentile
650 (raw score) =	1.5 (Z score) =	65 (T score) =	93rd percentile
660 (raw score) =	1.6 (Z score) =	66 (T score) =	95th percentile

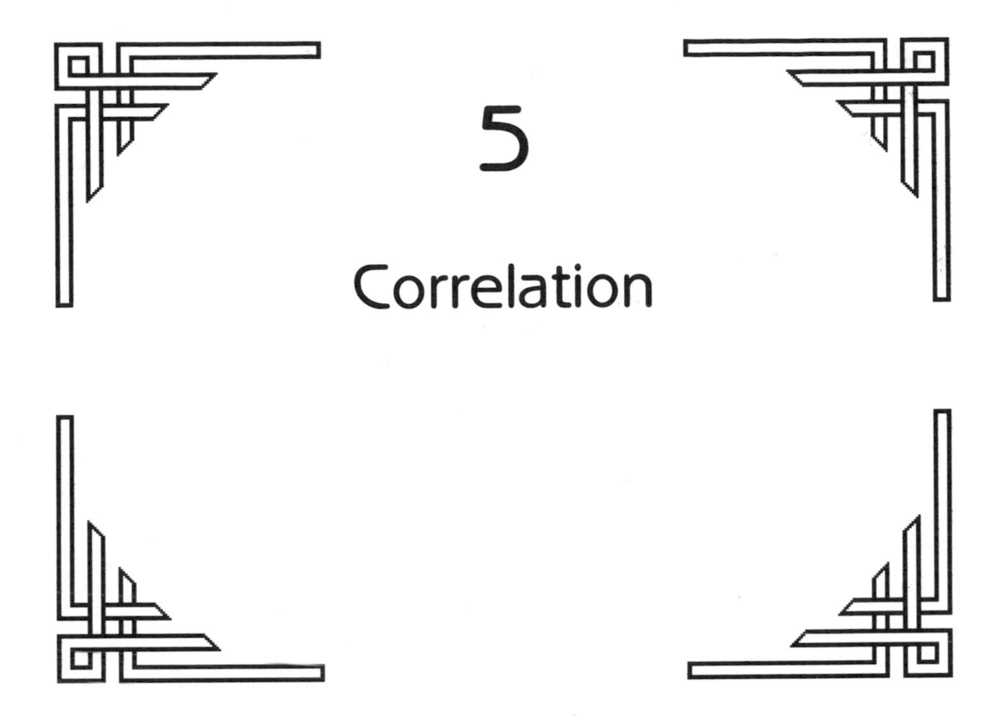

5 Correlation

Up to now we have been dealing with one variable at a time. In this chapter we will discuss how to measure the relationship between two variables. With such a measure we will be able to answer important questions such as: Is intelligence related to achievement? Is counselor empathy related to counseling outcome? Is student toenail length related to success in graduate school?

The relationship between two variables can be depicted by means of a scatter diagram. Suppose a group of students has taken an aptitude test. We can designate their Z scores as the X variable (we'll plot it on the X-axis of our diagram) and their T scores as the Y variable (plotted on the Y-axis). Each student has two scores. We *plot* a given student's position on the graph by drawing an invisible line out into the center of the graph from the vertical axis value (the Y value, in this case the student's T score) and drawing another invisible line up into the graph from the X-axis value (the student's Z score). Put a dot where the two lines cross, and you've plotted that student's location.

Student	Z Score	T Score
Alice	0	50
Bob	1	60
Corey	1.5	65
Darren	–.2	48
Evelyn	–.8	42

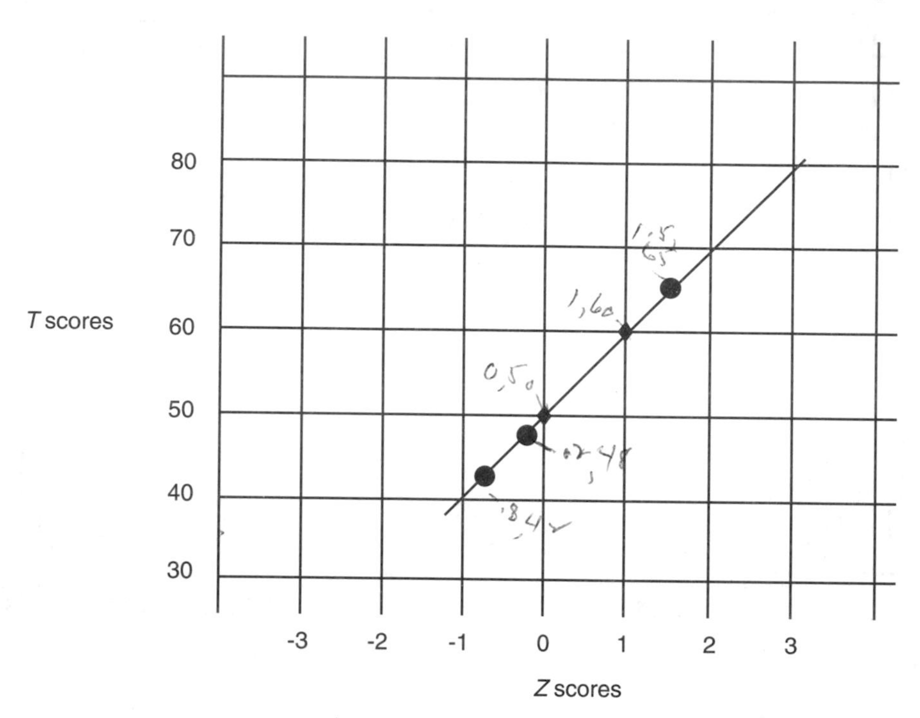

Figure 5-1 Relationship between *Z* scores and *T* scores.

As you can see in Figure 5-1, all the points representing the relationship between *Z* scores and *T* scores fall on a straight line. When two variables have this kind of relationship, we say that they are *perfectly correlated.* This means that if we know the value of something on one of the variables we can figure out exactly its value on the other. If we know that someone has a *T* score of 35, their *Z* score has to be –1.5.

Graphs that show relationships like this in picture form are useful, but often we need a more numerical way of expressing the same thing. In the next few pages, you'll learn how to compute a *correlation coefficient,* which is a quantitative measure of relationship. The relationship between *Z* scores and *T* scores would have a correlation of +1.00, a perfect positive relationship. Here are some facts about correlation coefficients:

1. The values of correlation coefficients range from –1.00 to +1.00, representing perfect negative and perfect positive correlations, respectively. A coefficient of 0.00 represents no relationship at all.[1]

[1]Not quite true, actually—a correlation coefficient of 0 means no *linear* relationship at all. Two variables could be *curvilinearly* related (we're going to talk about this later) and have a correlation coefficient of 0.

2. A positive correlation coefficient indicates that those individuals who scored high on one variable also tended to score high on the other (in our example, a high Z score means a high T score). When you plot the relationship between two positively correlated variables, the dots tend to fall along a line that runs from the lower-left corner of the graph to the upper-right corner.
3. A negative correlation indicates that when the value on one of the variables is high, it will be low on the other. As in, for example, number of cookies eaten in mid-afternoon (variable X) and amount of meat, potatoes, and spinach eaten at dinner (variable Y). Two negatively correlated variables tend to plot out along a line running from the upper-left corner of the graph to the lower-right corner.
4. When the correlation is either +1.00 or –1.00 (an unlikely event in educational or psychological research), all the points in the plot fall on a straight line. As the correlation gets farther from 1.00 (toward 0), the points *scatter* out away from the line. That's why such graphs are known as *scatter plots.*

Figure 5-2 on the next page shows sample scatter plots depicting various degrees of correlation.

THE PEARSON PRODUCT–MOMENT CORRELATION COEFFICIENT (r)

Many measures of correlation have been developed by statisticians for different purposes. The most common measure is the Pearson product–moment correlation coefficient, designated by lowercase r. Try to relax as much as you can before you look at the next computational formula, because it looks horrendous. Smile. We'll do it together, following the indicated steps, one at a time. It isn't as bad as it looks.

$$r = \frac{N\sum XY - (\sum X)(\sum Y)}{\sqrt{\left[N\sum X^2 - (\sum X)^2\right]\left[N\sum Y^2 - (\sum Y)^2\right]}}$$

Suppose you wanted to compute the correlation between intelligence test scores (designated as X) and the grade-point averages (Y) earned by a group of fifth graders:

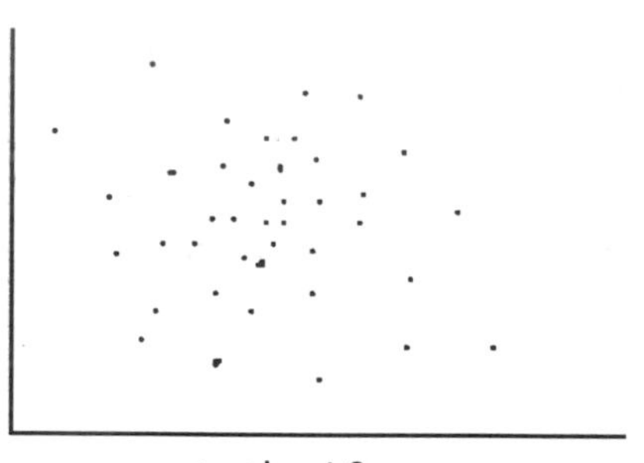

r = about 0

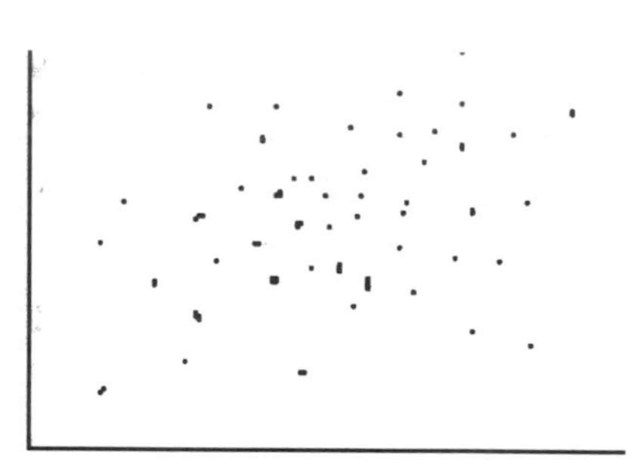

r = about .3

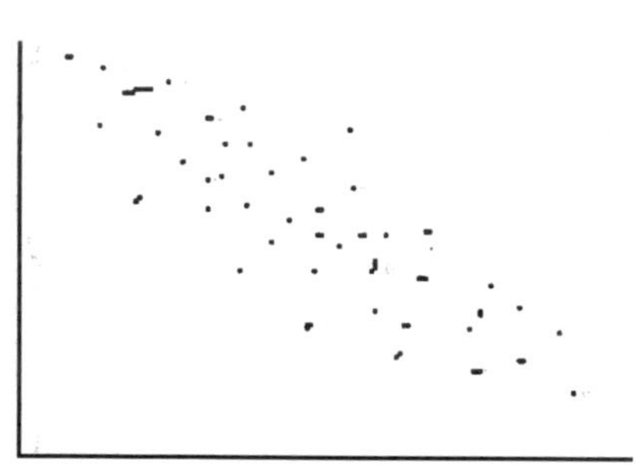

r = about -.8

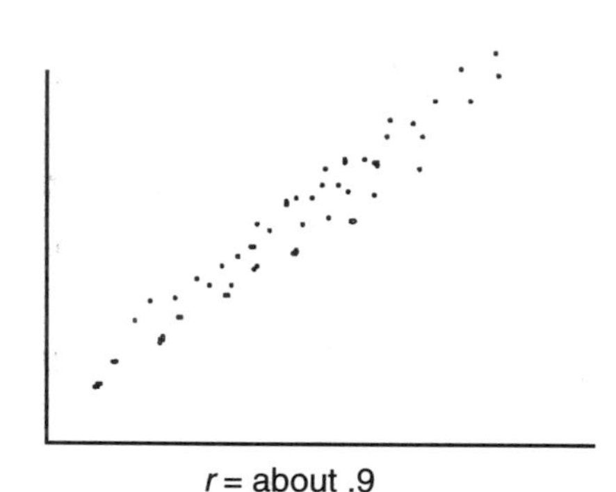

r = about .9

r = about 0
(This is an example of a curvilinear relationship, for which r is not the appropriate statistic.)

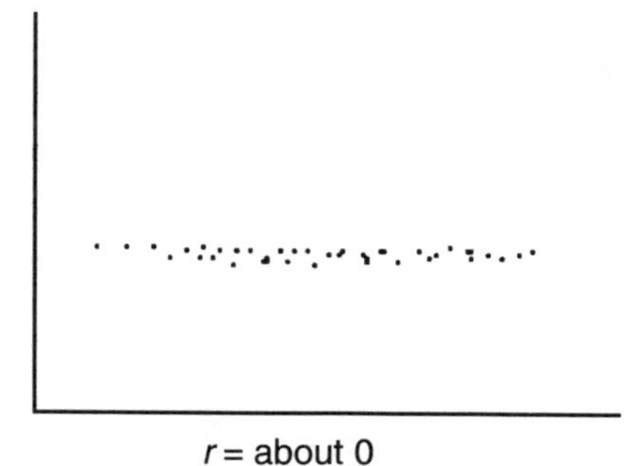

r = about 0
(Even though the points are more or less on a straight line, there is no relationship between being high on one variable and being either high or low on the other.)

Figure 5-2 Sample graphs depicting various degrees of correlation.

Intelligence Test Scores (X) and Grade-point Averages (Y) Earned by a Group of Fifth Graders

Student	IQ Score (X)	GPA (Y)
1	118	3.6
2	113	3.4
3	131	3.6
4	126	3.6
5	124	3.9
6	109	4.0
7	118	3.7
8	122	3.7
9	127	4.0
10	115	3.5

If you'll look closely at the formula for the Pearson product–moment correlation coefficient r, you'll notice that many of the values and computations are the same as for the variance. You might want to go back and review Chapter 2 now. Only one term in this formula is new: the $N\Sigma XY$ term found in the numerator. N is, of course, the number of subjects. In our example, $N = 10$. The rest of the term, ΣXY directs you, first, to multiply each person's X score by his or her Y score, and then to sum the products. I'll add an XY column to the data above so you can see what I mean:

Student	IQ Score (X)	GPA (Y)	IQ × GPA (XY)
1	118	3.6	424.8
2	113	3.4	384.2
3	131	3.6	471.6
4	126	3.6	453.6
5	124	3.9	483.6
6	109	4.0	436.0
7	118	3.7	436.6
8	122	3.7	451.4
9	127	4.0	508.0
10	115	3.5	402.5
	$\Sigma X = 1203$	$\Sigma Y = 37.0$	$\Sigma XY = 4452.3$

Now you have everything you need to compute r:

1. Look at the numerator of the formula for r first:
   a. Find $N\Sigma XY$. In our example, $N = 10$ and $\Sigma XY = 4452.3$.

$$N\Sigma XY = (10)(4452.3) = 44{,}523$$

b. Find $(\Sigma X)(\Sigma Y)$. $\Sigma X = 1203$ and $\Sigma Y = 37.0$.

$$(\Sigma X)(\Sigma Y) = (1203)(37.0) = 44{,}511$$

c. $N\Sigma XY - (\Sigma X)(\Sigma Y) = 44{,}523 - 44{,}511 = 12$.

2. Now let's look at the denominator:

a. Look inside the square root sign at the left-hand side of the denominator first. Notice that it is exactly the same as the numerator in the formula for the variance of X:

$$N\Sigma X^2 - (\Sigma X)^2 = (10)(145{,}149) - 1{,}447{,}209 = 4281$$

b. Now look at the right-hand side of the denominator and, again, notice that it is the same as the numerator for the variance of Y:

$$N\Sigma Y^2 - (\Sigma Y)^2 = (10)(137.28) - 1369 = 3.8$$

c. Now notice that you are directed to multiply the left-hand side of the denominator by the right-hand side:

$$[N\Sigma X^2 - (\Sigma X)^2][N\Sigma Y^2 - (\Sigma Y)^2] = [4281][3.8] = 16{,}267.8$$

d. Your next step is to calculate the final value of the denominator. To do this, take the square root of the value you just got in step 2c. The square root of 16,267.8 is 127.55.

3. You're (finally) ready to calculate r! All you have to do is divide the numerator (the value you found in step 1c) by the denominator (step 2d):

Numerator =12

Denominator =127.55

Dividing the numerator by the denominator gives you r:

$$r = \frac{12}{127.55} = .09$$

Before going on to discuss the meaning of the Pearson product–moment r just obtained, check your computational skills on the following data:

Student	5th Grade IQ (X)	9th Grade GPA (Y)
1	104	2.9
2	92	2.8
3	108	3.2
4	99	2.0
5	95	2.7
6	102	3.0
7	117	3.6
8	90	2.6
9	107	3.1
10	119	3.8
$N = 10$	$\Sigma X = 1033$	$\Sigma Y = 29.7$
	$\Sigma X^2 = 107{,}573$	$\Sigma Y^2 = 90.55$
	$\Sigma XY = 3104.4$	
	$r = .81$	

In our two examples of correlations between intelligence test scores (X) and GPA (Y), you obtained r =.09 and r =.81. What do these values mean?

There are a number of possible interpretations:

1. You will recall that a perfect correlation between two variables would result in $r = 1.00$, whereas, if there were no correlation at all between X and Y, then $r = 0$. Since both r's are positive and neither of the obtained r's equals either 1.00 or 0.00, we can say that there may be some correlation between the variables X and Y, but the relationship is not perfect.

2. How high are these correlations? Relative to what? Relative to the first correlation ($r = .09$), the second ($r = .88$) is quite high. Relative to correlations usually obtained between intelligence test scores and GPA, .09 is quite low, while .88 is a bit higher than usual ($r = .60$ would be more typical for this grade level). The point I'm trying to make here is that one way to judge whether or not an obtained r is high or low is to compare it with correlation coefficients typically obtained in similar studies. For example, if you have studied psychological measurement, you know that .60 as a predictive validity coefficient would be fairly high, but as a measure of split-half reliability it might be relatively low. (And if you haven't studied measurement and don't know what those big words mean, don't worry—this just wasn't a good example for you.)

3. It is important not to confuse correlation with causation. In the second example, $r = .81$; the two variables are fairly closely related. But that doesn't prove that intelligence causes achievement, any more than achievement causes intelligence. There probably is a positive correlation between the number of counselors and the number of alcoholics in a state, but one does not necessarily cause the other. Two variables may be correlated with each other due to the common influence of a third variable.

On the other hand, if two variables do not correlate with each other, one variable cannot be the cause of the other. If there is no difference in counseling outcome between, for example, "warm" and "cold" counselors, then "warmth" cannot be a factor determining counseling outcome. Thus, the correlational approach to research can also help us to rule out variables that are not likely to be important to our theory or practice.

This point is so important that it bears repeating: if two variables are correlated, it is still not necessarily true that one causes the other. However, if they are *not* correlated, one can*not* be the cause of the other.

4. The Pearson product–moment r is a measure of linear relationship between two variables. Some variables are related to each other in a curvilinear fashion. For example, the relationship between anxiety and some kinds of performance is such that low levels of anxiety facilitate performance, whereas high levels interfere with it. A scatter diagram showing the relationship between the two variables would show the dots distributing themselves along a curved line (curvilinear), rather than a straight line (linear). r is not an appropriate measure of relationship between two variables that are related to each other in a curvilinear fashion. If you are interested in the relationship between two variables measured over a large number of subjects, it's a good idea to construct a scatter diagram for 20 or so of these subjects before calculating r, just to get a general idea of whether the relationship is linear or curvilinear.

5. Another common way to use r is to calculate what is known as the coefficient of determination, which is the square of the correlation coefficient. For example, if the obtained r between intelligence test scores and GPA were $r = .60$, squaring would give you .36 (I know that it seems wrong when you see that the square of a number is smaller than the number being squared, but it's true. Try it.) The obtained square of the correlation coefficient (in our example, .36) indicates that 36% of the variability in Y is "accounted for" by the variability in X. "Why is it that some people achieve more in school than others?" some might ask. If intelligence were a cause of achievement and if the correlation between intelligence and achievement were $r = .60$, then 36% (squaring .60 and converting to percents) of the variability in achievement among people would be "caused" by intelligence, the other 64% being caused by other factors. Another way of saying the same thing is that if $r_{XY} = .60$, variables X and Y have 36% of their variability "in common."

6. When presenting statistical data in textbooks or professional journals, the author often will indicate whether or not the r obtained in the study was "significant" or "statistically significant." You will be exposed to an extended discussion of the concept of statistical significance when you study inferential statistics in later chapters in this book, but I will introduce one explanation now.

I own a pocket calculator that will compute r's at the push of a button. If I wish, I can sit and punch in random numbers representing fictitious X and Y variables, press the appropriate button, and get the correlation between X and Y. Even though the numbers entered were random, the obtained r is almost never exactly zero. Just by chance, the numbers usually come out as if there were some slight relationship among them. The point of this digression is that, when you compute r and it comes out to some number other than zero, how do you know it wasn't the result of "chance"?

Enter the concept of statistical significance. Perhaps you have read studies that said things like, "The obtained results are significant at the .05 or the 5% level." What the author is saying is that only 5 times out of 100 could you expect to get a result like the one obtained in the study by chance alone. If you and 999 other students sit and punch random numbers into your calculators and then press the "correlate" button, about 50 of you will get a value of r that is "significant at the .05 level." The rest of you will get lower values. About 10 of the 50 who got .05 significance will get a value big enough to be "significant at the .01 level," meaning that only 1 time in 100 would you get a result like that just by chance.

Fortunately for us, we don't have to do any mathematics to determine if our obtained r is statistically significant. Statisticians have done that for us. Turn now to Appendix C, the table of r values. To enter the table you will need first to locate the appropriate degrees of freedom (df) for your study. You'll learn more about df when you get to inferential statistics, too, but for now all you'll need to know is that $df = N - 2$, where N is the number of subjects in your study. Since $N = 10$ for both of the r values we computed, $df = N - 2 = 8$ for both.

Now that you know your df, move down the df column until you get to 8, and then look across. If you want to know whether or not your obtained r is significant at the .05 level, go to the column headed with .05. For df = 8, r's of .6319 or larger are significant at the .05 level. That is, an r as large as or larger than .6319 will turn up just by chance only 5 times in 100. The r of the first example ($r = .09$) does not meet this standard. In the second sample, though, the r ($r = .81$) is greater than .6319, and we can say that the probability of its occurring just by chance is less than 5% ($p < .05$).

Notice also that the correlation coefficient obtained in the second example ($r = .81$) is significant at the .01 level, indicating that you would expect to get an r that size or larger by chance alone only 1 time in 100 ($p < .01$).

If you had 20 subjects in your study ($N = 20$), then df = 18 and an obtained r of .4438 would be significant at the .05 level, and an r of .5751 would be significant at the .01 level. The more subjects (observations) in your sample, the less likely it is that you'll get a high value of r just by chance.

When your df falls between tabled values, use the values for the next lower df. For example, if $N = 60$ and df = 58, you would use the significance values for df = 50 (an $r = .2732$ would be significant at the .05 level).

There are a number other of correlational techniques in addition to the Pearson product–moment r. One of them, the Spearman correlation for ranked data, is presented later, along with a discussion of which measure of correlation is most appropriate for your particular data.

PROBLEMS

1. Given the following N's and obtained r's, indicate whether the result is significant at the .05 level:

a. $N = 100, r = .22$
b. $N = 100, r = .60$
c. $N = 9, r = -.70$
d. $N = 36, r = .23$
e. $N = 100, r = -.19$

2. Given the following data, what is the correlation between

a. IQ scores and anxiety test scores?
b. IQ scores and statistics exam scores?
c. anxiety test scores and statistics exam scores?

Indicate whether the obtained r is statistically signicant and, if so, at what level. (You may want to save your work on this one; we'll be using the same data for problems at the end of Chapter 6.)

Student	IQ (X)	Anxiety (Y)	Statistics Exam Scores (Z)
1	140	14	42
2	130	20	44
3	120	29	35
4	119	6	30
5	115	20	23
6	114	27	27
7	114	29	25
8	113	30	20
9	112	35	16
10	111	40	12

Answers

1. (a) significant; (b) significant; (c) significant; (d) n.s. (not significant); (e) n.s.

2. With $N = 10$ and df = 8, and $r_{crit\,\alpha=.05}$] = .55

$r_{XY} = -.61$; n.s. $r_{XZ} = .88$; significant $r_{YZ} = -.66$; significant

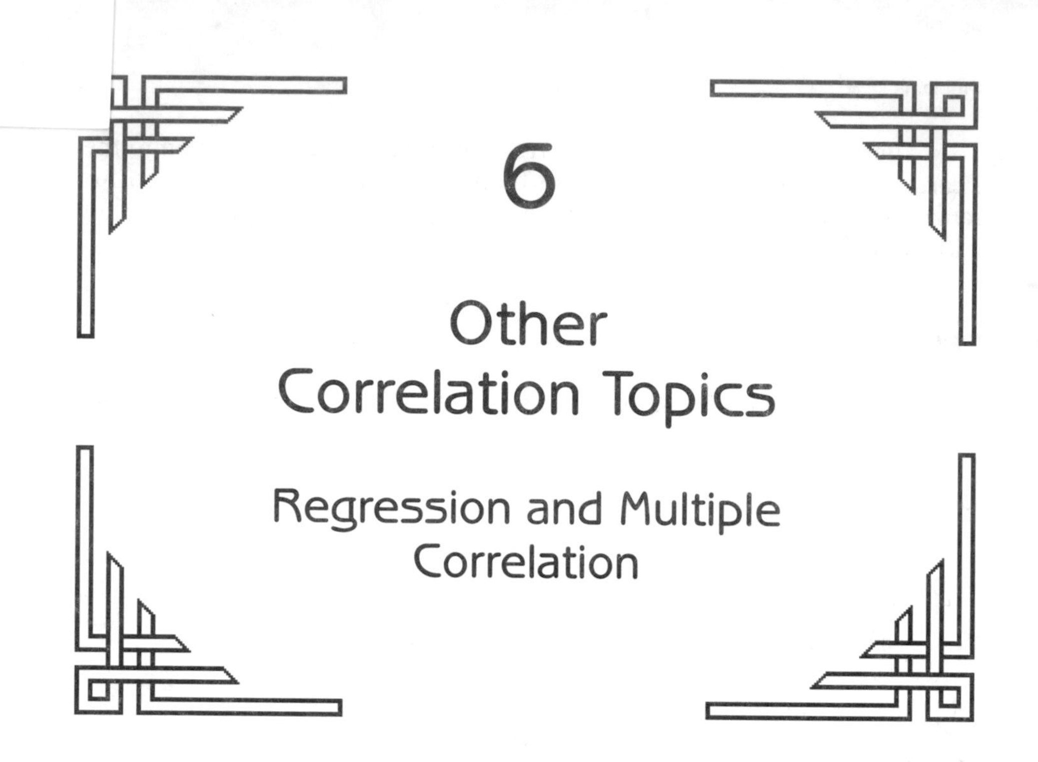

6

Other Correlation Topics

Regression and Multiple Correlation

REGRESSION

You know from Chapter 5 that when there is a positive correlation between two variables, X and Y, those persons who score high on the X variable also tend to score high on the Y variable, and those who score low on X tend to score low on Y. For example, suppose you know that there is a positive correlation between intelligence and achievement. Given a person's intelligence test score, you can predict his or her performance on an achievement measure. If you knew that John's IQ score was 140, you would predict a higher grade-point average for him than for Jim, whose IQ score was 60. You could increase the accuracy of your prediction considerably if you used what statisticians call a regression equation:

$$Y' = \left(r_{XY} \frac{S_Y}{S_X} X \right) - \left(r_{XY} \frac{S_Y}{S_X} \overline{X} \right) + \overline{Y}$$

Now, don't have a nosebleed; we'll take it apart:

- Y' is the predicted score on the Y variable, given a particular score X on the X variable. The prime (′) after the Y indicates that the computed value is an estimate of Y, rather than the true value of Y.
- r_{XY} is the Pearson product–moment correlation coefficient, showing

the relationship between X and Y variables (you learned how to compute r in Chapter 5).

- S_Y is the standard deviation of the scores on the Y variable, but it's computed using N·(N–1) in the denominator. I'll explain why in Chapter 9.
- S_X is the standard deviation of the scores on the X variable, with the same change in the denominator.
- X is the score earned by an individual on the X variable, the predictor variable. We're trying to predict that person's score on the Y variable.
- $\overline{X}$ is the mean of the X variable
- $\overline{Y}$ is the mean of the Y variable

Let's work it out with an example. Given next are the ninth-grade Differential Aptitude Test (DAT) scores and subsequent grade-point averages (GPA) earned by a sample of 18 senior high school students:

Student	DAT Score (X)	GPA (Y)
1	45	2.7
2	64	3.2
3	44	2.6
4	22	2.6
5	33	2.9
6	24	3.1
7	30	1.5
8	29	2.0
9	15	1.5
10	64	3.8
11	45	2.4
12	43	2.7
13	34	1.8
14	40	1.7
15	19	1.3
16	25	2.5
17	33	1.9
18	29	1.6

$\Sigma X = 638$ $\quad \Sigma Y = 41$

$\Sigma X^2 = 25{,}814$ $\quad \Sigma Y^2 = 101.45$

$\overline{X} = 35.44$ $\quad \overline{Y} = 2.28$

$S_X = 13.72$ $\quad S_Y = .67$

$\Sigma XY = 1569.50$

$r = .72$

$$Y' = \left(.72\frac{.67}{13.72}X\right) - \left(.72\frac{.67}{13.72}35.44\right) + 2.28$$

$$= .035X - (.035 \cdot 35.44) + 2.28$$

$$= .035X + 1.03$$

That's very nice—but it would be even nicer if it had some usefulness. And it does! We can use this regression equation to estimate or predict Sam's grade-point average once we know his DAT score. Let's say he scored 36 on the DAT:

$$Y'_{\text{SAM}} = .035(36) + 1.03 = 2.29$$

Suzie's DAT score is 75. What's her predicted GPA?

$$Y'_{\text{SUZIE}} = .035(75) + 1.03 = 3.36$$

Notice that the basic formula is always the same. The only thing that changes is the X value plugged in for each person.

Now, do these numbers mean that Sam will have a GPA of exactly 2.29, and Suzie's will be exactly 3.36? No, but their GPAs will probably be close to those values. Wait a minute—what's this "probably" stuff, and how close is "close"? These are legitimate questions, and our next job is to answer them. Unfortunately, the answer isn't simple and will require some explanation.

STANDARD ERROR OF THE ESTIMATE

Schools and employers use a variety of measures to predict how well people will do in their settings. While such predictions usually are more accurate than those made by nonmathematical methods (guesstimation), there is some danger in using the results of predictions in a mechanical way.

Suppose, for example, that the data we've been looking at, the DAT and grade-point averages, didn't come from ninth graders but rather from a college admissions test and college GPA data. Suppose also that Denny scores 25 on the admissions test. His predicted GPA is

$$Y'_{\text{DENNY}} = .035(25) + 1.03 = 1.90$$

Since his predicted GPA is less than a C average, it might be tempting to conclude that Denny is not "college material." Before jumping to that conclusion, however, it is wise to remember that not everyone who has a predicted GPA of 1.90 achieves exactly as predicted: some do better than predicted (overachievers?), while others do worse (underachievers?).

In fact, if you took 100 persons, all of whom had a predicted GPA of 1.90, their actual GPAs would vary considerably. You could compute a mean and standard deviation of the actual GPAs. Theoretically, the actual scores would be distributed normally and would have a mean equal to the predicted GPA of 1.90, and the standard deviation would be

$$S_{\text{est}} = S_Y\sqrt{1 - r_{XY}^2}$$

where S_{est} is called the standard error of the estimate

r^2 is the squared correlation between the X and Y variables

Now this is a very important idea. We're talking about a group of people selected out of probably thousands who took the college admissions test—selected because that test predicted that they would have a 1.9 GPA. But they *don't* all have a GPA of 1.9: some did better than predicted, and some not so well. Their GPAs cluster around the predicted value. If there were enough of them (it would probably take more than 100 in order to work out perfectly), the distribution would be normal in form, would have a mean equal to the predicted value, and would have a standard deviation of $S_{est} = .67\sqrt{1 - .72^2} = .46$.

You will recall from Chapter 3 that approximately 68% of the scores in a normal distribution fall between plus and minus one standard deviation from the mean. In this example, the standard deviation of our distribution of GPA scores (S_{est}, the standard error of the estimate) is .46. So 68% of the scores in this distribution can be expected to fall between 1.90 – .46, and 1.90 + .46, or between 1.44 and 2.36. In other words, 68% of the scores fall between ±1 standard error of the estimate from the mean.

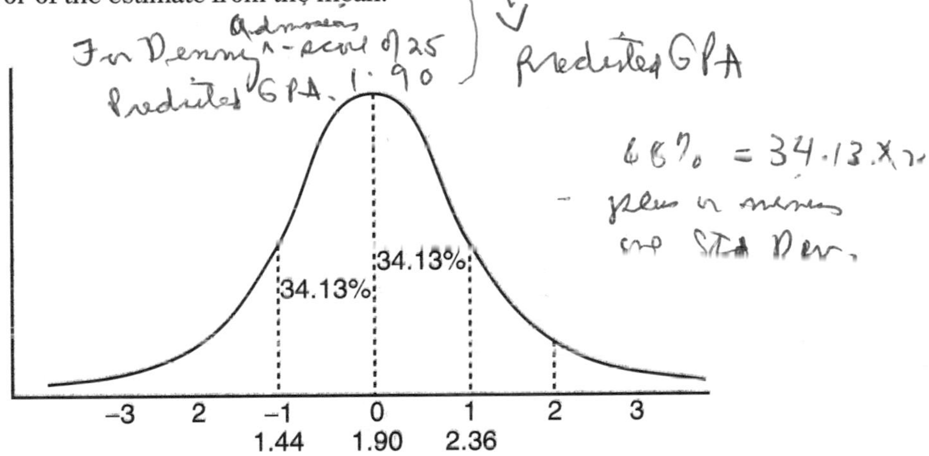

If you define someone who graduates from college with a C average as "college material," then you can see that a fairly good percentage of those who were predicted to have a 1.90 GPA (less than a C average) actually would graduate (that is, have an actual GPA higher than 2.0). In fact, using the table in Appendix B, you can figure out exactly how many students are likely to fall into this category:

1. A GPA of 2.0 is .10 above the mean (2.0 – 1.90).
2. That's .1/.46 of a standard deviation, or .22 standard deviations above the mean.
3. Going to the table, we find that .09 of the distribution lies between the mean and .22 sd above the mean. Since 50% of the distribution is above the mean, 50% – 9% = 41% of the GPAs will be higher than 2.0.

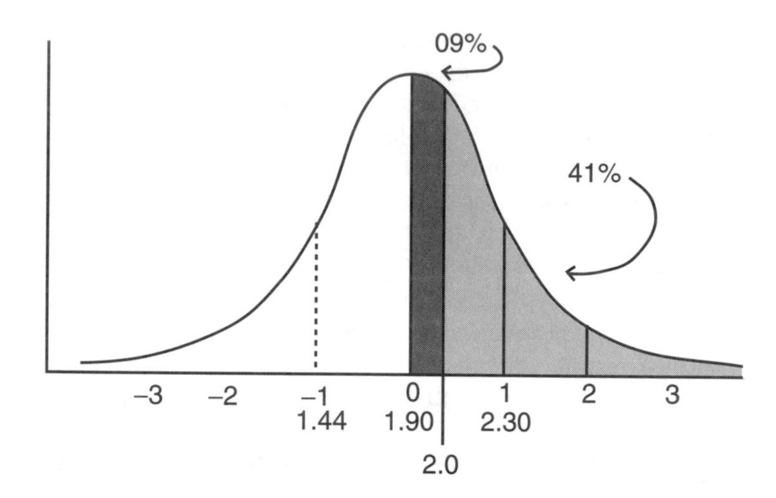

As I said before, it's very logical if you just take it one step at a time. About 41% of the total population of students would be mislabeled if we used a cutting point of exactly 1.90.

Look at the formula for the standard error of the estimate again:

$$S_{\text{est}} = S_Y\sqrt{1 - r_{XY}^2}$$

Notice that, if $r = 1.00$, $S_{est} = 0$, which is another way of saying that, if the correlation between X and Y were perfect, there would be no errors in predicting performance on Y from our predictor X. Unfortunately (or fortunately?), there are virtually no perfect correlations between predictor and predicted variables in measuring human beings. We are not able to predict human performance with anything near perfection at this time—nor do I believe that we ever will.

MULTIPLE CORRELATION

Suppose you wished to predict the grade-point average in your college (call this variable the Z variable), using *two* predictors, college freshman IQ test scores (X) and high school GPA (Y). Suppose also that 18 subjects scored as follows:

Subject	IQ (X)	High School GPA (Y)	College GPA (Z)
1	114	3.1	2.7
2	130	3.1	3.2
3	117	3.1	2.6
4	109	1.9	2.6
5	121	3.4	2.9
6	100	3.1	2.4
7	109	2.0	1.5
8	115	2.7	2.0
9	107	1.1	1.5
10	138	3.7	3.8
11	124	3.4	2.4
12	130	3.1	2.7
13	116	2.4	1.8
14	120	2.6	1.7
15	122	3.1	1.3
16	115	3.6	2.5
17	116	2.6	1.9
18	105	1.8	1.6
	$\overline{X} = 117.11$	$\overline{Y} = 2.80$	$\overline{Z} = 2.28$
	$S_X = 9.57$	$S_Y = .67$	$S_Z = .67$
	$r_{XY} = .60$	$r_{YZ} = .61$	$r_{XZ} = .58$

At the bottom of the table you can see the correlation between IQ scores (X) and high school GPA (Y), $r_{XY} = .60$; between high school GPA (Y) and college GPA (Z), $r_{YZ} = .61$; and between IQ (X) and college GPA (Z), $r_{XZ} = .58$. What we want to do is combine our two predictors, high school GPA and IQ scores, in the mathematically best way, and then find out how well this combination correlates with college GPA. The result of all this calculation is known as a multiple R. In this case it will be designated $R_{Z.XY}$, which indicates that X and Y are predictors and Z is the predicted variable.

Theoretically, you can combine as many predictor variables as you want in order to predict the value of some other variable. At one time some psychologists believed that if, in addition to variables like IQ and high school GPA, they added other variables as predictors, such as motivation to learn, interest, and personality factors, each new predictor would improve the prediction. Eventually, they thought, they would be able to achieve multiple *R*'s approaching 1.0. So far, they haven't been able to do this; the improvement in prediction tends to drop very rapidly as each new predictor is added.

Here is the formula and a worked-out example for multiple *R* based on the preceding data: two predictor variables, IQ and high school grades, predicting a third variable, college GPA. Note that the formula yields R^2, not R. You will need to take the square root of R^2 to end up with $R_{Z.XY}$.

$$R^2_{Z.XY} = \frac{r^2_{XZ} + r^2_{YZ} - 2(r_{XZ})(r_{YZ})(r_{XY})}{1 - r^2_{XY}}$$

In our example

$$R^2_{Z.XY} = \frac{.58^2 + .61^2 - 2(.58)(.61)(.60)}{1 - .60^2} = .45;\ R_{Z.XY} = .67$$

As you can see, when you combined *X* and *Y* predictors to get a multiple *R*, the correlation was slightly higher than obtained by using either predictor alone. This value can be interpreted in much the same way as a Pearson product–moment *r*, except that you enter Appendix C with $n - 3$ degrees of freedom, rather than $n - 2$. Methods are available for using three or more variables as predictors, but you would do best to consult with a friendly computer person if you want to use them.

PARTIAL CORRELATION

Suppose that you were interested in the effect of counselors' touching their clients during counseling sessions. You observed a large number of counselors and counted the number of times each counselor touched his or her client (call this variable *X*); at the end of the session, you asked each client to fill out a client satisfaction measure (call this variable *Y*). And you got a significant positive correlation—hurrah! But when you suggested to your fellow counselors that your data indicated that frequent counselor touch contributed to greater client satisfaction, a skeptic argued that perhaps the correlation was not due to counselor touch at all, but rather to counselor empathy. The skeptic went on to say that she believed that more empathic counselors touch more—clients like those counselors not because they touch, but because they are empathic and understanding. The correlation between satisfaction and touch is just an irrelevant side effect of the relationship between touching and empathy.

To answer this criticism, you might well choose to use the statistical technique known as ***partial correlation.*** Partial correlation allows you to measure the degree of relationship between touch (variable X) and client satisfaction (variable Y) with the effect of empathy (variable Z) "partialed out" or "controlled for."

To do partial correlation, you need a way of measuring each variable you're concerned with. You've already measured client satisfaction and touching; let's say you also ask independent raters to observe the counselors at work and rate them for "degree of empathy."

You've collected your data on all three variables and have found that the correlation between counselor touch (X) and client satisfaction (Y) is $r_{XY} = .36$; the correlation of empathy (Z) and counselor touch (X) is $r_{XZ} = .65$; and the relationship between empathy (Z) and client satisfaction (Y) is $r_{YZ} = .60$. The formula for finding the relationship between counselor touch (X) and client satisfaction (Y), with counselor empathy (Z) partialed out is

$$r_{XY.Z} = \frac{r_{XY} - (r_{XZ})(r_{YZ})}{(1 - r_{XZ}^2)(1 - r_{YZ}^2)}$$

Now we can use this formula to compute the correlation of touch and client satisfaction, with empathy "controlled for":

$$r_{XY.Z} = \frac{.36 - (.65)(.60)}{\left[1 - (.65)^2\right]\left[1 - (.60)^2\right]} = .05$$

This result would tend to support the hypothesis of your skeptical friend: there doesn't seem to be a very strong relationship, if any, between counselor touch and client satisfaction when empathy is partialed out.

PROBLEMS

1. Using the data from the problem set at the end of the last chapter, find the regression equation that would be used to predict:
   a. anxiety scores, if you know the IQ scores.
   b. statistics test scores, if you know the IQ scores.
   c. statistics test scores, if you know the anxiety scores.
2. Given these regression equations, what would you predict as an anxiety score for a person who scored 125 on the IQ test? What about that person's statistics score? What statistics score would you predict for someone with an anxiety score of 20?
3. Compute the multiple correlation for the IQ–anxiety–statistics data, considering IQ and anxiety as predictors of statistics scores.
4. A personnel direction of a large computer firm gave all its employees "quickie" IQ

tests as well as questionnaires to measure their job satisfaction. She reported that job satisfaction was clearly related to IQ, since the two measures had a correlation of .42. A company executive pointed out, however, that the correlation might really have more to do with salary than with IQ—that smart people tend to be paid more, and people who get more money tend to like their jobs better. Given the following values, compute the correlation between IQ and job satisfaction, with salary effects partialed out: $r_{\text{IQ, satisfaction}} = .42$; $r_{\text{IQ, salary}} = .46$; $r_{\text{salary, satisfaction}} = .63$

Answers

1. **a.** Anxiety = $-.63_{\text{IQ}} + 99.84$ **b.** Statistics = $.99_{\text{IQ}} - 90.2$ **c.** Statistics = $-.69_{\text{anxiety}}$ $+ 44.65$
2. **a.** 33.55 **b.** 21.09 **c.** 30.85
3. $R_{\text{statistics,IQ,anxiety}} = .81$; $R_{\text{statistics,IQ,anxiety}} = .91$
4. $r_{XY.Z} = .23$

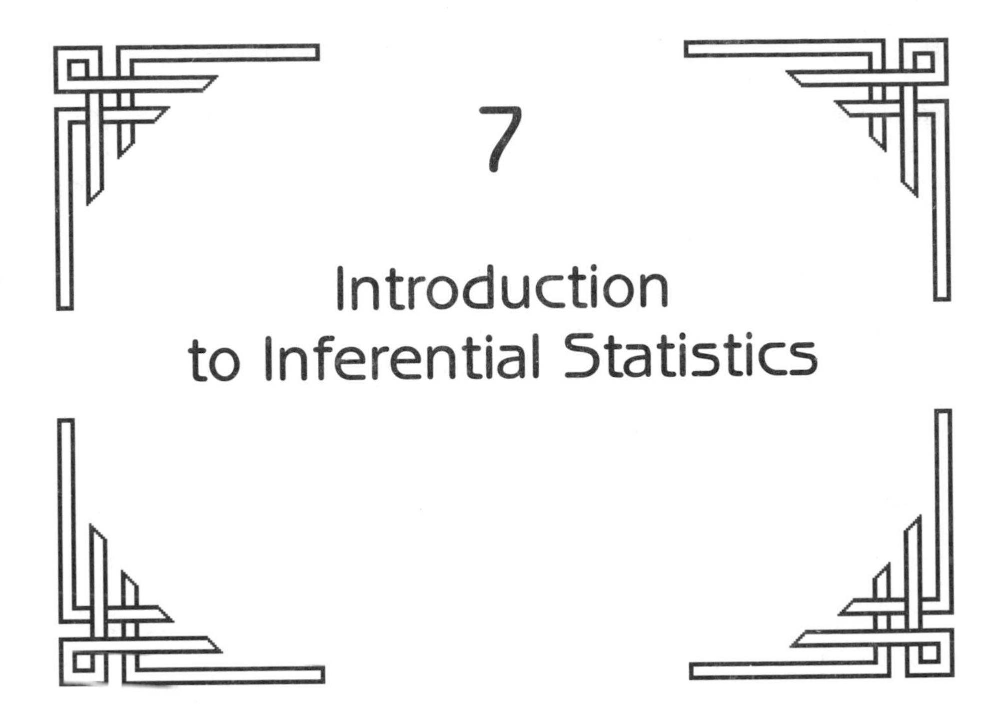

7

Introduction to Inferential Statistics

At harvest time, wheat farmers take their loads of grain to market in large trucks. Any particular load contains mostly wheat, but it usually also contains a percentage of weed seeds. The buyer of the wheat takes a small sample of the grain, analyzes the sample carefully to find what percentage is wheat and what percentage weeds, and then pays the farmer for the percentage of the entire truck load that is estimated (on the basis of the sample) to be wheat. Similarly, American pollsters study small samples of people in order to make predictions about how an entire country will vote at election time. It works the same way: use a small, easily measurable sample to estimate some characteristic of the whole population that the sample was drawn from.

Suppose you wished to compare the math anxiety of male and female graduate students in the United States. You could, theoretically, administer a math anxiety test to all female and male graduate students in the country, score the tests, and confidently draw a conclusion about whether males or females were more math anxious. Chances are good, however, that your resources would not allow you to conduct such a large study.

Because it usually is not possible, for financial or other reasons, to study entire populations, it is necessary to study samples from those populations. If you wish to generalize what you learn about your samples back to the populations from which they were drawn, your major concern must be that your samples not be biased. (Perhaps the most notorious biased sample was taken inadvertently in 1936, when the *Literary Digest* predicted on the basis of its sample of automobile and telephone owners that Alf Landon, the Republican candidate, would win the presidential election by a landslide. In 1936, many people didn't have telephones or cars, and the ones who did have them were more likely to be Republicans than Democrats.)

Insofar as possible, your sample should be representative of the population in all respects. But that is easier said than done! In fact, all samples probably contain some error (this is known as *sampling error*). For example, suppose that a large jar contained 1000 marbles, 500 red and 500 black. You mix the marbles up as well as you can and then draw a blind sample of 100 marbles. Would you get exactly 50 red and 50 black marbles? Probably not. Now imagine taking repeated samples of 100 marbles, dumping each sample back in and remixing after you examine it. By chance alone, some of the samples would have more black than red marbles, and vice versa. Most of your samples would have nearly an even split, but, by chance or sampling error alone, a few samples might be quite different: one might have 70 of one color and only 30 of the other, another might be divided 60–40, and so on.

Although all methods of sampling will inevitably result in some degree of error, some methods are more likely to give you representative samples than others. In this book we will concentrate on one method of sampling—*random* sampling—because it is one of the best and most widely applicable methods.

There are two basic principles in random sampling: (1) each individual in a population must have an equal chance of being selected, and (2) the selection of one individual must be independent of the selection of any other individual. A good way to ensure that both of these rules are followed is to use a table of random numbers. Appendix D provides such a table, along with instructions for using it to select a random sample.

THE NULL HYPOTHESIS

Ordinarily, researchers are interested in demonstrating the truth of some hypothesis of interest: that some relationship exists between two variables; that two groups differ in some important way; that population X is bigger, stronger, smarter, more anxious, and so on, than population Y. We need statistical procedures to test such hypotheses. Problem is, though, that it's almost impossible (with most statistical techniques) to demonstrate that something is *true*. Statistical techniques are much better at demonstrating that a particular hypothesis or statement is *false*, that it's very unlikely that the hypothesis could really hold up.

So we have an interesting dilemma. We want to show that something is true, but our best tools only know how to show that something is false. The solution is both logical and elegant: state the exact opposite of what we want to demonstrate to be true, disprove that, and what's left—what hasn't been disproved—must be true.

In case you are now thoroughly confused, here's an example. A researcher wants to show that boys, in general, have larger feet than girls. (I know, it's a dumb idea, but grants have been awarded for stranger things.) She knows that her statistical tools can't be used to demonstrate the truth of her hypothesis. So she constructs what's known as a *null hypothesis,* which takes in every possibility except the one thing she wants to prove: boys' feet are either smaller or just the same size as girls' feet. If she can use her statistical techniques to disprove or reject this null hypothesis, then there's only one thing left to believe about boys' and girls' feet—the very thing she wanted to prove in the first place.

The hypothesis that the scientist wants to support or prove is known as the *research hypothesis;* the "everything else" hypothesis is called the *null hypothesis* and is symbolized as H_0. A primary use of inferential statistics is that of attempting to reject H_0.

If we could measure everyone in the population(s) of interest, we wouldn't need to use statistics; we could just look at the measurements and say, "Yes, my hypothesis is true" or "No, I must have been wrong." Measurements of whole populations (rare though they are) are known as *parameters.* In contrast, measures of samples are called *statistics.* The mean of a whole population is symbolized as μ; it's a parameter. The mean of a sample is symbolized $\overline{X}$; it's a statistic. Statisticians use statistics to make inferences about the populations from which the samples were drawn.

Getting back to our study of the math anxiety of male versus female graduate students, H_0 is that there is no difference between the mean anxiety scores of male and female graduate student populations; that is, H_0: $\mu_{\text{Females}} = \mu_{\text{Males}}$. The purpose of our study is to decide whether H_0 is probably true or probably false.

Suppose we drew 50 subjects (using a random sampling method, of course) from each of the two populations, male graduate students and female graduate students, and tested them for math anxiety. Suppose also that the mean of our female sample was 60 and the mean for our male sample was 50:

$$\overline{X}_{\text{Males}} = 50, \qquad \overline{X}_{\text{Females}} = 60$$

Obviously, the mean of our female sample was higher than that for males. Does this prove that the null hypothesis is not true?

There are two possible explanations for our observed difference between male and female sample means: (1) there is, in fact, a difference in math anxiety between the male and female population means, that is, $\mu_{\text{Female}} \neq \mu_{\text{Male}}$, and the difference we see between the samples reflects this fact; or (2) there is no appreciable difference in anxiety between the means of the male and female graduate student populations, that is, $\mu_{\text{Female}} = \mu_{\text{Male}}$, and the difference we observe between the sample means is due to chance or sampling error (this would be analogous to drawing more black marbles than red even though there is no difference in the proportion of the two in the jar from which the sample was drawn).

If the null hypothesis really is true, then the differences we observe between sample means is due to chance. The statistical tests you will study in later chapters, such as the *t* test or analysis of variance, will help you to decide if your obtained results may have been due to chance or if there probably is a difference in the two populations. If the result of your study is that the difference you observe is "statistically significant," then you will reject the null hypotheses and conclude that you believe there is a real difference in the two populations.

TYPE I AND TYPE II ERRORS

Your decision either to reject or not to reject the null hypothesis is subject to error. Because you have not studied all members of both populations and because statistics (such as the sample mean) are subject to sampling error, you can never be completely sure whether H_0 is true or not. In drawing your conclusion about H_0, you can make two kinds of errors, known (cleverly) as Type I and Type II errors. Rejection of a true null hypothesis is a Type I error, and failure to reject a false null hypothesis is a Type II error. Perhaps the table at the top of the next page will help you to understand the difference.

If the real situation is that there is no difference in math anxiety between males and females, but you reject H_0, then you have made a Type I error; if you do not reject H_0, then you have made a correct decision. On the other hand, if there is a real difference between the population means of males and females and you reject H_0, you have made a correct decision, but if you fail to reject H_0, you have made a Type II error.

Investigator's Decision	The Real Situation (unknown to the investigator)	
	H_0 is true $\mu_{Females} = \mu_{Males}$	H_0 is false $\mu_{Females} \neq \mu_{Males}$
Reject H_0	Investigator makes a Type I error	Investigator makes a correct decision
Do not reject H_0	Investigator makes a correct decision	Investigator makes a Type II error

The primary purpose of inferential statistics is to help you to decide whether or not to reject the null hypothesis and to estimate the probability of a Type I or Type II error when making your decision. Inferential statistics can't tell you for sure whether or not you've made either a Type I or a Type II error, but they can tell you how likely it is that you have made either type.

One last point: notice that you do not have the option of *accepting* the null hypothesis. That would amount to using your statistical test to "prove" that the null hypothesis is true, and you can't do that. Your two possible decisions really amount to either (1) I reject the null hypothesis, and so I believe that there really are important differences between the two populations, or (2) I can't reject the null hypothesis, and I still don't know whether there are important differences or not. For this reason, Type I errors are generally considered to be more serious than Type II errors. Claiming to have gotten significant results when there really are no differences between the populations is a more serious mistake than saying that you don't know for sure, even when those differences might exist. As you will see, statistical decisions are usually made so as to minimize the likelihood of a Type I error, even at the risk of making lots of Type II errors.

STATISTICAL SIGNIFICANCE AND THE TYPE I ERROR

Suppose that a colleague of yours actually did the study of math anxiety that we've been talking about and concluded that the difference between male and female sample means was "statistically significant at the .05 (or the 5%) level." This statement would mean that the difference that he observed between the sample means could have occurred only 5 times out of 100 by chance alone. Since it could have happened only 5 times out of 100 just by chance, your colleague may be willing to bet that there is a real difference in the populations of male and female graduate students and will reject the null hypothesis.

You must realize, however, that whenever you reject the null hypothesis you may be making an error. Perhaps the null hypothesis really is true, and this is one of those 5 times out of 100 when, by chance alone, you got this large a difference in sample means. Another way of saying the same thing is that, if the null hypothesis were true, 5 times out of 100 you would make a

Type I error when you use this decision rule. You would reject the null hypothesis when it was, in fact, true 5% of the time.

You might say, "But, I don't want to make errors! Why can't I use the .01 (or 1%) level of significance instead of the .05 level? That way, I reduce the

Type I error

Type II error

likelihood of a Type I error to 1 out of 100 times." You can do that, of course, but when you do so, you increase the probability of a Type II error (more about that later). Suffice it to say here that, in educational and psychological research, it is conventional to set the .05 level of significance as a minimum standard for the rejection of the null hypothesis. Typically, if an obtained result is significant at, say, the .08 level, an author will conclude that he or she was unable to reject the null hypothesis or that the results were not statistically significant.

Making a Type II error, failing to reject the null hypothesis when it's really not true, is rather like missing your plane at the airport. You didn't do the right thing this time, but you can still catch another plane. Making a Type I error is like getting on the wrong plane—not only did you miss the right one, but now you're headed in the wrong direction! With a result significant at the .08 level, the odds still suggest that the null hypothesis is false (you'd only get this large a difference by chance 8 times in 100). But researchers don't want to get on the wrong airplane; they'd rather make a Type II error and wait for another chance to reject H_0 in some future study.

PROBLEMS

1. Which of the following would *not* be a truly random sample and why?
   a. To get a sample of children attending a particular grade school, the researcher numbered all the children on the playground at recess time and used a random number table to select 50 of them.
   b. Another researcher, at a different school, got a list of all the families who had kids in that school. She wrote each name on a slip of paper, mixed up the slips, and drew out a name. All kids with that last name went into the sample; she kept this up until she had 50 kids.
   c. A third researcher took a ruler to the school office, where student files were kept. She used a random number table to get a number, measured that distance into the files, and the child whose file she was over at that point was selected for the sample. She did this 50 times and selected 50 kids.
2. What is the appropriate H_0 for each of the following research situations?
   a. A study to investigate possible differences in academic achievement between right- and left-handed children
   b. A study to determine if registered nurses give a different level of patient care than LPNs
   c. A study exploring whether dogs that were raised in kennels have different training patterns than dogs raised in homes
3. Assume that each of the following statements is in error: each describes a researcher's conclusions, but the researcher is mistaken. Indicate whether the error is Type I or Type II.
   a. "The data indicate that there are significant differences between males and females in their ability to perform task 1."

b. "There are no significant differences between males and females in their ability to perform task 2."
c. "On the basis of our data, we reject the null hypothesis."
d. "On the basis of our data, we cannot reject the null hypothesis."

4. a. Explain, in words, the meaning of the following: "The difference between group 1 and group 2 is significant at the .05 level."
   b. When would a researcher be likely to use the .01 level of significance rather than the .05 level? What is the drawback of using the .01 level?

Answers

1. a. Not random because all members of the population didn't have an equal chance of being included (kids who were ill or stayed inside during recess couldn't be chosen).
   b. Not random because selection wasn't independent (once a given child was selected, his or her siblings were included, too).
   c. Not random, because kids who had very thick files (which usually means they caused some sort of problem) would have a better chance of being selected).
2. **a.** $\mu_{\text{left}} = \mu_{\text{right}}$; **b.** $\mu_{\text{RN}} = \mu_{\text{LPN}}$; **c.** $\mu_{\text{kennel}} = \mu_{\text{home}}$
3. **a.** Type I; **b.** Type II; **c.** Type I; **d.** Type II
4. a. If we performed this experiment over and over and if the null hypothesis were true, we could expect to get these results just by chance only 5 times out of 100.
   b. We use the .01 level when we need to be *very* sure that we are not making a Type I error. The drawback is that as we reduce the probability of a Type I error, the likelihood of a Type II error goes up.

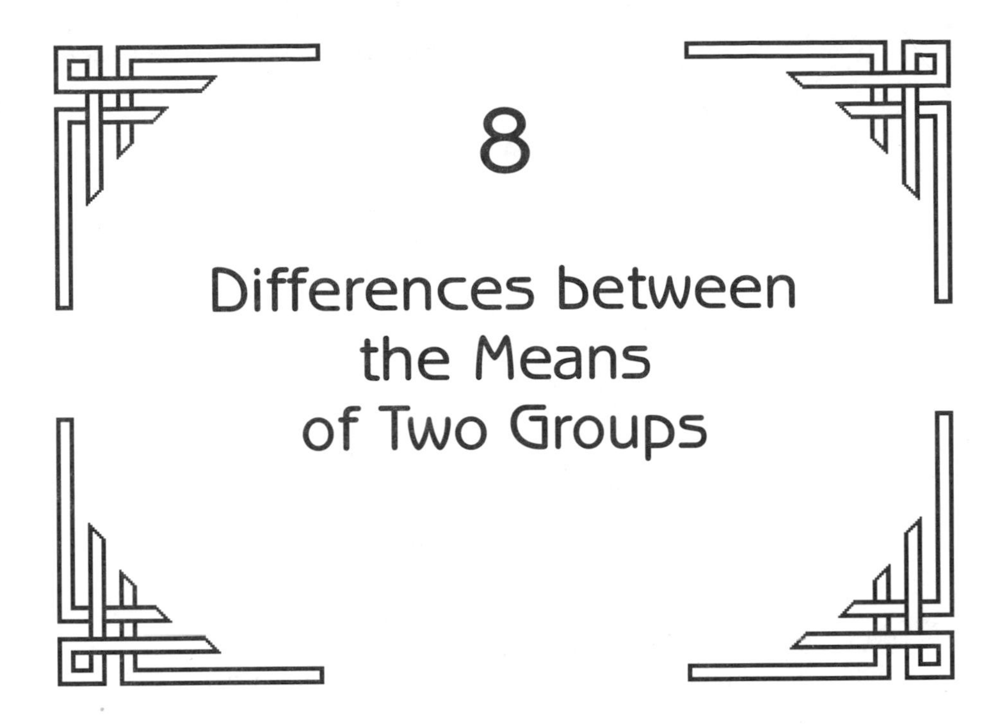

8 Differences between the Means of Two Groups

THE *t* TEST

The *t* test is one of the most commonly used inferential statistics. Its primary purpose is to determine whether the means of two groups of scores differ to a statistically significant degree. This is the kind of question we were asking in Chapter 7, when we looked at math anxiety levels of male and female graduate students. Here's another example: suppose that you randomly assigned 12 subjects each to group 1, a counseling group, and to group 2, a waiting-list control group. Suppose also that after those in group 1 had been counseled you administered a measure of psychological adjustment to the two groups, with results as shown on the next page.

The hypothesis you will test is the null hypothesis:

$$H_0: \mu_1 = \mu_2$$

The null hypothesis states that there is no difference in mean adjustment level between those who receive counseling and those who don't. As you can see, there is a difference in the two sample means, but it may be that these observed differences occurred by chance or sampling error. We need to find out if the difference is statistically significant. If the difference between $\overline{X}_1$ and $\overline{X}_2$ is statistically significant, you will reject the null hypothesis and conclude that there is a difference in adjustment level between people who have had counseling and those who have not.

Counseled Group Scores	Control Group Scores
25	14
14	11
23	13
21	9
24	15
17	12
19	9
20	11
15	8
22	13
16	12
21	14
$\overline{X}_1 = 19.75$	$\overline{X}_2 = 11.75$
$n_1 = 12$	$n_2 = 12$
$S_1^2 = 12.93$	$S_2^2 = 4.93$

Since the counseled and control groups in our hypothetical study were independent of each other, we would use the *t* test for independent samples. Here we go!

THE *t* TEST FOR INDEPENDENT SAMPLES

The *t* test, like most other statistical tests, consists of a set of mathematical procedures that yields a numerical value. In the case of the *t* test, the larger the value, the more likely it is to reflect a significant difference between the two groups under comparison. We'll learn how to compute the value of *t* in the next few pages; first, though, let's think about what that value really means.

A major application of the *t* test for independent samples is found in experimental research like the study in our example. Researchers often draw a sample from one population and randomly assign half of the subjects to an experimental and the other half to a control group, or to some other comparison group (see Appendix D for a method of assigning subjects at random to two or more groups). Since the subjects were assigned to their two groups by chance (that is, at random), the means of the two groups should not differ from each other at the beginning of the experiment any more than would be expected on the basis of chance alone. If a *t* test were done at the beginning of the experiment, the difference between the means would probably not be statistically significant.[1] After the experiment, the means of the two groups are compared using the *t* test. If the obtained value of *t* is large enough to be sta-

[1]If we are to be very precise here, we would say that the difference between the means would be significant at the .05 level only 5 times out of 100 or would be significant at the .01 level only once in 100 such experiments.

tistically significant (usually at the .05, or 5% level), the experimenter rejects H_0. Since the two groups have now been shown to differ more than would be expected on the basis of chance alone, and since the only difference between them (that we know of) is the experimental treatment, it is reasonable to conclude that this treatment is responsible for the differences we have observed.

The formulas for the t test look pretty horrendous at first. Just remember to relax and work on one step at a time. When you look closely, you will see that you learned how to do most of the computations in earlier chapters.

FORMULAS FOR THE t TEST FOR TWO INDEPENDENT SAMPLES

$$t_{obt} = \frac{\overline{X}_1 - \overline{X}_2}{S_{\overline{X}_1 - \overline{X}_2}}$$

$$S_{\overline{X}_1 - \overline{X}_2} = \sqrt{\frac{S_p^2}{n_1} + \frac{S_p^2}{n_2}}$$

$$S_p^2 = \frac{(n_1 - 1)S_1^2 + (n_2 - 1)S_2^2}{n_1 + n_2 - 2}$$

In these formulas,

t_{obt} is the value of t obtained through your data.

n_1 and n_2 are the number of subjects in each of the two groups.

S_1^2 and S_2^2 are the variances of the two groups computed using N·(N 1) in the denominator.

$\overline{X}_1$ and $\overline{X}_2$ are the means of the two groups.

Next, I have given you a worked out example using the data of the counseling and no-counseling groups. You might want to see if you can do it on your own before you look at the example. Notice that you have to compute the bottom-most formula first and then work your way up through the three formulas until you can solve for t_{obt}. And don't get discouraged if you can't do it on the first try, because I'll go through the computations with you a step at a time.

COMPUTATION OF t_{obt} FOR COUNSELED AND CONTROL GROUPS

Each successive formula of the three formulas given above provides some value to plug into the earlier one. Therefore, we begin at the bottom, with the formula for S_p^2.

Notice that this computation requires that you know the number of subjects (n) and the variance (S^2) for each of your groups. If you go back to the beginning of this chapter, you will see that both n_1 and $n_2 = 12$, $S_1^2 = 12.93$, and $S_2^2 = 4.93$. If you were doing your own study, you would, of course, have to compute both variances. But you know how to do that, right?

Go ahead now and do the arithmetic indicated in the formula (remember that multiplying and dividing are done before adding and subtracting, unless parentheses indicate otherwise). Your answer should come out to 8.93.

$$S_p^2 = \frac{(n_1 - 1)S_1^2 + (n_2 - 1)S_2^2}{n_1 + n_2 - 2} = \frac{[(12-1) \times 12.93] + [(12-1) \times 4.93]}{12 + 12 - 2} = 8.93$$

Next comes the middle formula. Use the value that you got in the first computation, 8.93, to plug into this one. Unless you make an arithmetic error, you'll get 1.22 as your answer.

$$S_{\overline{X}_1 - \overline{X}_2} = \sqrt{\frac{S_p^2}{n_1} + \frac{S_p^2}{n_2}} = \sqrt{\frac{8.93}{12} + \frac{8.93}{12}} = \sqrt{1.49} = 1.22$$

Finally, you're ready for the last step, the formula that gives the value of t_{obt}. Again, in your own study you would have to do some preliminary work, this time to get the means of the two samples. In our example, this has been done for you: $\overline{X}_1 = 19.75$ and $\overline{X}_2 = 11.75$. Note that these values go into the numerator of the formula. The value of t_{obt} is 6.56.

$$t_{obt} = \frac{\overline{X}_1 - \overline{X}_2}{S_{\overline{X}_1 - \overline{X}_2}} = \frac{19.75 - 11.75}{1.22} = 6.56$$

Notice, by the way, that the decision of which sample mean is subtracted from the other is purely arbitrary; we could just as well have used $\overline{X}_2 - \overline{X}_1$ for the numerator of that last equation. Had we done so, the value of t_{obt} would have been negative rather than positive. When the direction of the difference we are interested in is unimportant, the t test is *nondirectional* and we use the absolute value of t_{obt}: with a negative value, we would just drop the negative sign and proceed as if we had subtracted in the other direction.

THE CRITICAL VALUE OF *T*: t_{crit}

"Okay," you might say, "I've done all the computations. Now what does my t_{obt} mean?" Good question. To find out whether your t_{obt} is statistically significant—that is, if it is large enough so that it probably reflects more than chance or random differences between the two groups—you will have to compare it with what is known as the critical value of t (I'll designate the critical value of t as t_{crit}). To find t_{crit}, go to Appendix E. Look at the left-hand set of values (labeled Two-tailed or Nondirectional Test) and notice that the farthest column to the left is headed with the letters df.[2] The abbreviation df means "degrees of freedom." To find the degrees of freedom for a t test for independent samples, just subtract 2 from the total number of subjects in your study. In our example, $df = n_1 + n_2 - 2 = 12 + 12 - 2 = 22$.

t_{crit} is determined by df and by your selected level of significance. Suppose you selected the .05 level of significance (the most commonly chosen value). Go down the df column to 22, the number of df in our example, and across to the value in the column to the right (headed with .05*). There you will see the number 2.074. That is the critical value of t, or t_{crit}, for the .05 level of significance when df = 22. If your t_{obt} is equal to or greater than t_{crit}, your results are statistically significant at the .05 level. Another way of saying this is that there are fewer than 5 chances out of 100 that a t value this large could have occurred by chance alone, symbolized as $p < .05$.

The t_{obt} in our example was 6.56. This is obviously larger than $t_{crit} = 2.074$; therefore, your results are significant at the .05 level. In fact, if you go across the df = 22 row, you will see that your $t_{obt} = 6.56$ is greater than the t_{crit} for the .01 level (2.819) and for the .001 level (3.792). This means that your results are significant at the .001 level ($p < .001$), indicating that you could have gotten results like this less than 1 time out of 1000 by chance alone. You would conclude that differences between counseled and control groups favored the counseled group to a statistically significant degree.

Let's do one more example, going through each step in a typical study comparing two groups. Suppose you wanted to test the hypothesis that male and female counselors differ in the degree of empathy that they show their clients. You would select representative samples of male and female counselors and use some sort of test to measure their empathy. Imagine that your results were as follows (the higher the score, the greater the degree of empathy shown):

Group 1 Male Empathy Scores	Group 2 Female Empathy Scores
7	7
5	8
3	10
4	7
1	

Step 1: State Your Hypothesis. The statistical hypothesis you will be testing is the null hypothesis. In this example the null hypothesis is that there is no difference between populations of male and female counselors in

[2]Don't worry too much about what "two-tailed" means; we'll get back to it after you've learned how to do this first kind of t test.

the level of empathy that they offer their clients. In statistical terms,

$$H_0: \mu_M - \mu_F = 0$$

Sometimes hypotheses will be stated as alternative or research hypotheses, which are the opposite of the null hypotheses. In this case, the alternative hypothesis would be that there *is* a difference in populations of male and female counselors in the degree of empathy that they offer their clients,

$$H_1: \mu_M - \mu_F \neq 0$$

Step 2: Select α, Your Significance Level. The level of significance chosen is known as α (alpha). The most typical significance level selected is $\alpha = .05$. As indicated in Chapter 7, you might choose $\alpha = .01$ if you want to be even more careful to avoid committing a Type I error. If a significant result would commit you to investing a great deal of money in program changes, or would lead to other important policy decisions, for example, then a Type I error would be quite dangerous and you would want to be very cautious indeed in setting your α level.

Step 3: Compute t_{obt}

a. Find n, X, and S^2 for each group

Group 1 (males): $n_1 = 5, \overline{X} = 4, S^2 = 4$

Group 2 (females): $n_2 = 4, \overline{X} = 8, S^2 = 1.5$

b. Plug these values into the formulas:

$$S_p^2 = \frac{(n_1 - 1)S_1^2 + (n_2 - 1)S_2^2}{n_1 + n_2 - 2} = \frac{[(5-1) \times 4] + [(4-1) \times 1.5}{(5+4) - 2} = 2.93$$

$$S_{\overline{X}_1 - \overline{X}_2} = \sqrt{\frac{S_p^2}{n_1} + \frac{S_p^2}{n_2}} = \sqrt{\frac{(2.93)}{5} + \frac{(2.93)}{4}} = \sqrt{1.32} = 1.15$$

$$t_{obt} = \frac{|\overline{X}_1 - \overline{X}_2|}{S_{\overline{X}_1 - \overline{X}_2}} = \frac{|4 - 8|}{1.15} = 3.48$$

Step 4: Find t_{crit}. Entering Appendix E with df $= n_1 + n_2 - 2 = 4 + 5 - 2 = 7$, and with $\alpha = .05$, you can see that $t_{crit} = 2.365$.

Step 5: Decide Whether or Not to Reject the Null Hypothesis. As I pointed out earlier, the sign of t_{obt} will depend on which sample mean you happened to label $\overline{X}_1$ and which one you labeled $\overline{X}_2$. For the hypothesis you are testing now, it doesn't really matter which sample mean is larger; you're only interested in whether or not they're different. For this reason, use the absolute value of t_{obt}: if you get a negative value for t_{obt}, just change the sign to positive before comparing it to the tabled value. If t_{obt} has an absolute value equal to or greater than t_{crit}, you will reject H_0. Comparing $t_{obt} = 3.48$ with $t_{crit} = 2.365$, we decide to reject the null hypothesis. Our results are significant at the .05 level. In fact, our results almost reach significance at the .001 level. It is reasonable to conclude on the basis of our study that female counselors offer a higher level of empathy to their clients than do male counselors.

Requirement for Using the *t* Test for Independent Samples

The *t* test you have just learned requires that you have two independent samples, which means that the subjects for one group were selected independently from those in the second group. That is, the measurements from the two groups aren't paired in any way; a given measurement from group 1 doesn't "go with" a particular measurement from group 2. Sometimes you want to do a study in which this is not the case; for paired data you will use the *t* test for nonindependent groups.

THE *t* TEST FOR NONINDEPENDENT (MATCHED) SAMPLES

Suppose you gave a group of 10 subjects a test both before and after a movie intended to influence attitudes toward public schools. You had two sets of scores, one from the pretest and the other from the posttest, and you wanted to find out if attitudes as measured by the tests were more or less favorable after seeing the movie than they were before. (Notice that you now have pairs of scores, a score for each subject on the pretest and another score from each subject for the posttest. You have two groups of scores, but they are not independent of each other; they are matched.)

TABLE 8-1

Subject	Pretest Scores	Posttest Scores	Pretest – Posttest = D	D^2
1	84	89	+5	25
2	87	92	+5	25
3	87	98	+11	121
4	90	95	+5	25
5	90	95	+5	25
6	90	95	+5	25
7	90	95	+5	25
8	93	92	–1	1
9	93	98	+5	25
10	96	101	+5	25
	$\bar{X}_{pre} = 90$	$\bar{X}_{post} = 95$	$\Sigma D = 50$ $\bar{D} = 5$	$\Sigma D^2 = 322$

The results are shown in Table 8-1 (pretest and posttest scores are in columns 2 and 3, column 4 is the difference between each pair of scores, and column 5 is the square of that difference).

The most important thing to notice about these data, when deciding what statistical test to use, is that the scores are *paired.* The score of 84 goes with the score of 89, and it is the fact that subject 1 raised his or her score by 5 points that is important, rather than the values of the two scores by themselves. It wouldn't make sense to scramble the posttest scores and then look at the D (the difference between pretest and posttest) scores. Each pretest score is logically linked to one, and only one, posttest score. That's the definition of nonindependent samples; whenever that condition holds, then a nonindependent samples test is appropriate.

You will probably be pleased to discover that this nonindependent samples t test is easier to compute than the independent samples one. Here we go, step by step:

Step 1: State Your Hypothesis. Your null hypothesis is that there is no difference between attitudes before the movie and attitudes after the movie; that is,

$$H_0: \mu_{\text{pre}} - \mu_{\text{post}} = 0$$

Your alternative, or research, hypothesis is that there is a difference between attitudes before and after the movie:

$$H_1: \mu_{\text{pre}} - \mu_{\text{post}} \neq 0$$

Step 2: Select Your Level of Significance. For this example, we arbitrarily select the .05 level of significance.

Step 3: Compute t_{obt}. The formula is

$$t_{obt} = \frac{\overline{D}}{\sqrt{\frac{N \Sigma D^2 - (\Sigma D)^2}{N^2(N-1)}}} = \frac{5}{\sqrt{\frac{10(322) - (50)^2}{10^2(10-1)}}} = \frac{5}{.89} = 5.61$$

ΣD is the sum of the D column, the column of differences between pre- and posttest scores. In our example, $\Sigma D = 50$. Note that here we *do* pay attention to the sign of the differences; that -1 has to be subtracted, not added. Note also that the formula calls for $(\Sigma D)^2$, which is not the same as ΣD^2. Remember? In our example, $(\Sigma D)^2 = (50)^2 = 2500$. Finally, notice that I subtracted each subject's pretest score from his or her posttest score. I could have subtracted posttest from pretest scores, in which case most of my D scores would have been negative and my t_{obt} would also have been negative. It wouldn't have made any difference, however, as long as I did it the same way for every pair. If the absolute value of t_{obt} is greater than t_{crit}, it is statistically significant.

$\overline{D}$ is the mean difference, computed by dividing ΣD by N. In our example, $\overline{D} = 5$.

ΣD^2 is the sum of the D^2 column. In our example, $\Sigma D^2 = 322$.

N is the number of *pairs* of scores. In our example, $N = 10$. Don't confuse N with n, the number of subjects in a sample!

Step 4: Find t_{crit}. As was true for the t test for independent samples, we enter the table in Appendix E to find t_{crit}. In the case of the t for nonindependent samples, however, df $= N - 1$, where N is the number of subjects (the number of pairs of scores, not the total number of scores). Thus, in our example, df = 10 – 1 = 9 and t_{crit} = 2.262 at the .05 level of significance.

Step 5: Decide Whether to Reject H_0. As was the case with independent *t* tests, we compare our t_{obt} with t_{crit}. If t_{obt} (either positive or negative) is equal to or greater than t_{crit}, then we reject H_0 and conclude that our results are significant at the chosen level of α. Since in our example t_{obt} is greater than t_{crit}, we can reject the null hypothesis and conclude that the movie appeared to have a positive effect on attitudes toward public schools.

DIRECTIONAL VERSUS NONDIRECTIONAL TESTS

When you as a researcher are quite confident, on the basis of previous research or theory, that the mean of one group should be higher than that of some other group, and you predict the direction of the difference before you collect your data, you can then use what is called a one-tailed, or directional, *t* test. When conducting a one-tailed test, the t_{obt} formulas will be the same, but the value of t_{crit} will be different.

Suppose you predicted in advance that the mean adjustment score of a counseled group (n = 12) would be higher than the mean of a noncounseled control group (n = 12). Suppose also that t_{obt} = 2.0. If you were conducting a two-tailed test, t_{crit} with 22 degrees of freedom (do you know why df = 22? If

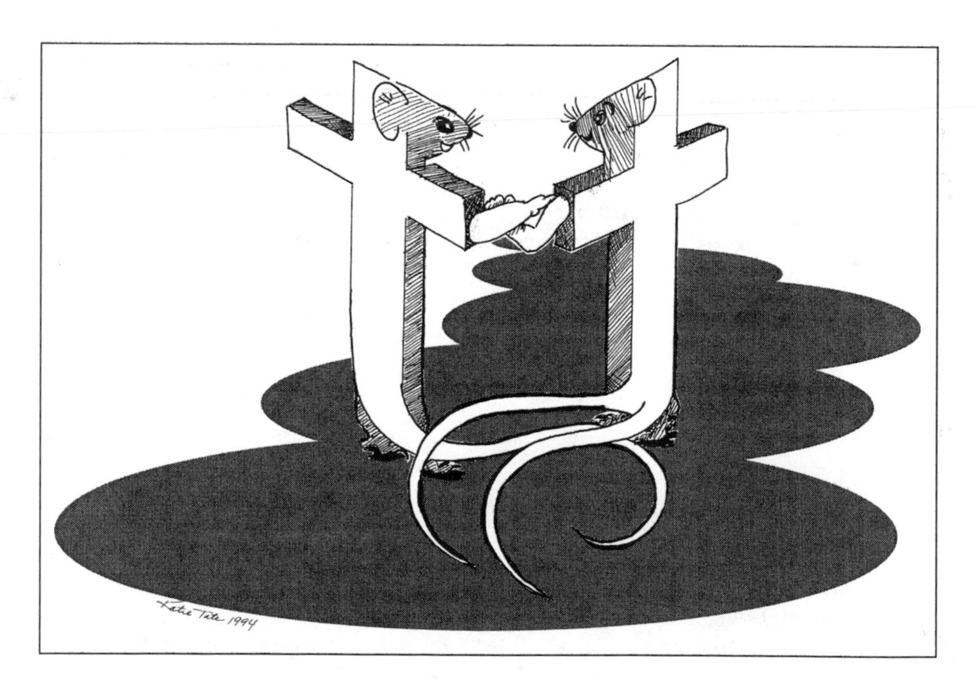

not, go back to page 79) would be 2.074 at the .05 level of significance. Since t_{obt} is smaller than t_{crit}, the result is not statistically significant.

Since, however, you predicted the direction of the difference in advance of your study, you can use a one-tailed test and use the t_{crit} values from the right-hand side of Appendix E. As you can see from that table, with df = 20, t_{crit} = 1.717 for a one-tailed test. t_{obt} is still 2.0, but now your results are statistically significant at the .05 level.

The diagrams at the top of Appendix E show how a value of t_{obt} can be significant for a one-tailed test but not for a two-tailed test. In a one-tailed test, we are only interested in differences in one direction—out in one tail of the distribution. If a t_{obt} is large enough to fall beyond the value of t_{crit} in that tail, t_{obt} is significant. With a two-tailed test, differences in *either* direction are important, and so we have to split the value of α and put half into each tail of the distribution. With α = .05, t_{crit} will mark off .025 at either end. Because the area marked off is smaller, t_{crit} must "slide" farther out away from 0, the mean; as it slides away from the mean, it gets bigger. With everything else equal, the value of t_{crit} will always be larger for a two-tailed test than for a one-tailed test. It's easier to get significant results with a one-tailed test, because t_{obt} doesn't have to be so large.

So why not always use a one-tailed test? To use the one tailed test legitimately, you must make your predictions prior to data collection. To do otherwise would be analogous to placing your bets after you see the outcome of an event. When in doubt, it is better to do two tailed tests, if only to avoid temptation. However, doing a two-tailed test does increase the likelihood of a Type I error, that is, of not rejecting the null hypothesis when it should be rejected. If a significant outcome of your research would make sense *only* if the observed differences are in a particular direction (if you'd dismiss anything else as chance or random differences, no matter how unlikely), then do a one-tailed test. Remember, though, that if you choose to do a one-tailed test and your data show "significant" differences in the opposite direction, you may not reject H_0. By choosing a one-tailed approach you have committed yourself to assuming that *any* differences in the non-predicted direction are due to chance or error.

PROBLEMS

For each test, be sure to specify the null hypothesis being tested, and whether you will use a *t* test for independent samples or a *t* test for nonindependent samples; also, specify whether you will use a one-tailed or two-tailed test.

1. In a study designed to discover whether men or women drink more coffee, a researcher (working on a *very* limited budget) observes five men and five women randomly selected from her university department. Here's what she found:

Number of Cups of Coffee in 1 Day at Work	
Men	Women
5	8
1	3
4	7
2	3
3	5

a. Run the appropriate test, assuming that both men and women were originally part of one random sample, with $n = 10$, and were then divided into men's and women's groups.

b. Use the same data, but this time assume that the men were randomly selected and then women were selected so that each man could be matched with a woman of the same age, job classification, and overall health status.

2. Using hospital and agency records, you locate six pairs of identical twins, one of whom was adopted at birth and the other of whom was in foster care for at least 3 years. All the twins are now 5 years old. You want to show that early adoption leads to better intellectual adjustment, so you test all the twins with the Wechsler Intelligence Scale for Children (WISC). Your results:

	WISC Score	
Twin Pair No.	Adopted Twin	Foster-care Twin
1	105	103
2	99	97
3	112	105
4	101	99
5	124	104
6	100	110

3. The following table contains test scores for three groups of clients at a college counseling center. Group 1 clients have received six sessions of counseling; group 2 clients were put on a waiting list for 6 weeks and asked to keep a personal journal during that time; group 3 clients were put on the waiting list with no other instructions. Use a t test to decide whether:

a. group 2 (journal) clients scored differently from group 3 (control) clients.

b. group 1 (counseled) clients scored differently from group 2 (journal) clients.

Group 1 (Counseled)	Group 2 (Journal)	Group 3 (Control)
22	6	8
16	10	6
17	13	4
18	13	5
	8	2
	4	

d. group 1 (counseled) clients scored higher than group 3 (control) clients.

4. A researcher tests the high-frequency hearing acuity of a group of teens two days before they attend a rock concert; 2 days after the concert she tests them again. Here are her results; she hopes to show that the teens hear better before the concert than afterward (the ***higher*** the score on this test, the poorer the hearing).

Subject	Scores (Preconcert)	Scores (Postconcert)
Tom	12	18
Dan	2	3
Sue	6	5
Terri	13	10
Karen	10	15
Lance	10	15
Christy	5	6
Jan	2	9
Lenora	7	7
Roberta	9	9
Dave	10	11
Victoria	14	13

Answers

1. a. H_0: $\overline{X}_{Men} = \overline{X}_{Women}$
   t test for independent samples; two-tailed test
   $S_p = 3.85$; $S^2\overline{X}_{Men} - \overline{X}_{Women} = 1.24$
   $t_{obt} = 1.77$; df $= 8$; $t_{crit} = 2.306$; do not reject H_0
   b. H_0: $\overline{X}_{Men} = \overline{X}_{Women}$
   t test for nonindependent samples; two-tailed test
   $t_{obt} = 5.88$; df $= 4$; $t_{crit} = 2.776$; reject H_0
2. H_0: $X_{Adopted} < \overline{X}_{Foster}$
   t test for nonindependent samples; one-tailed test
   $t_{obt} = .966$; df $= 5$; $t_{crit} = 2.571$; do not reject H_0
3. a. H_0: $\overline{X}_2 = \overline{X}_3$
   t test for independent samples; two-tailed test
   $t_{obt} = 2.12$; df $= 9$; $t_{crit} = 2.262$; do not reject H_0
   b. H_0: $\overline{X}_1 = \overline{X}_2$
   t test for independent samples; two-tailed test
   $t_{obt} = 4.0$; df $= 8$; $t_{crit} = 2.306$; reject H_0
   c. H_0: $\overline{X}_1 > \overline{X}_2$
   t test for independent samples; one-tailed test
   $t_{obt} = 8.2$; df $= 7$; $t_{crit} = 1.895$; reject H_0
4. H_0: $\overline{X}_{Pre} < \overline{X}_{Post}$
   t test for nonindependent samples; one-tailed test
   $t_{obt} = 1.898$; df $= 11$; $t_{crit} = 1.796$; reject H_0

9 One-way Analysis of Variance

In Chapter 8 you learned how to determine if the means of two groups differ to a statistically significant degree. In this chapter you will learn how to test for differences among the means of three or more groups.

Suppose you assigned subjects to one of three groups: a counseled group (group 1) an exercise/diet group (group 2), and a no treatment control group (group 3), with posttreatment adjustment scores as follows:

Group 1 (Counseled)	Group 2 (Exercise/Diet)	Group 3 (Control)
22	6	8
16	10	6
17	13	4
18	13	5
	8	2
	4	
$\overline{X} = 18.25$	$\overline{X} = 9$	$\overline{X} = 5$

You could test for the differences among these means with the *t* test; and you could test for significance of difference for X_1 versus X_2, X_1 versus X_3, and X_2 versus X_3. Here are two reasons why it would not be a good idea to do this kind of analysis:

1. It's tedious. If you do t tests, you will have to compute $k(k-1)/2$ of them, where k is the number of groups. [In our example, $3(3-1)/2 = 3$. This isn't too bad, but if you were comparing means among, say, 10 groups, you would have to compute $10(10-1)/2 = 45$ t tests!]
2. More importantly, when you select, for example, the .05 level of significance, you expect to make a Type I error 5 times out of 100 by sampling error alone. If you performed 20 t tests and one of them reached the .05 level of significance, would that be a chance occurrence or not? What if 3 of the 20 were significant—which would be chance and which would be a "real" difference? With multiple t's, you really don't know what the probability of a Type I error is, but it almost surely is greater than .05.

When you have three or more groups in your study, usually you will want to use an analysis of variance (ANOVA) as the way to analyze for statistical significance of the differences among the means of the groups. It may be important to note here that, even though the name of this statistic has the term "variance" in it, it is used to test for significant difference among *means*. The test looks at the amount of variability (the differences) between the means of the group, compared with the amount of variability among the individual scores within each group—that is, the variance between groups versus the variance within groups—and that's where the name comes from. The ANOVA starts with the total amount of variability in the data and divides it up (the statisticians call it "partitioning") into various categories. Eventually, the technique allows us to compare the variability among the group means with the variability that occurred just by chance or error—and that's exactly what we need to be able to do.

Perhaps you recall the definitional formula for the variance, given to you in Chapter 2:

$$S^2 = \frac{\sum\left(X - \overline{X}\right)^2}{N}$$

This is the formula for the variance of a single sample. When we are using a sample value to estimate the variance of the population from which the sample was drawn, this formula doesn't work. It turns out that, other things being equal, that the smaller the n of any group, the smaller its variance. Since samples are usually much smaller than the populations they represent, variances of samples will tend to be systematically smaller than the variances of their populations. The solution is simple: we enlarge the sample variance. If we use the degrees of freedom $(n-1)$ instead of n, the denominator becomes slightly smaller and the value of the variance grows correspondingly larger.[1]

[1]That's why N–1 showed up in the t test and the regression computations too.

$$S^2 = \frac{\sum(X - \overline{X})^2}{n-1}$$

Degrees of freedom is a concept you may not be familiar with (if you are, you can skip this paragraph). The basic idea has to do with the number of scores in a group of scores that are free to vary. In a group of 10 scores that sum up to 100, you could let 9 of the scores be anything you wanted. Once you had decided what those 9 scores were, the value of the tenth score would be determined. Let's say we made the first 9 scores each equal to 2. They'd add up to a total of 18; if the sum has to be 100, then the tenth score has to be 82. The group of 10 scores has only 9 degrees of freedom, 9 scores that are free to vary, 9 df.

The first step in carrying out an analysis of variance is to compute the variance of the total number of subjects in the study—we put them all together, regardless of the group that they've been assigned to, and find the variance of the whole thing. We do this using $n_T - 1$ (the total degrees of freedom) for the denominator of the formula:

$$S_T^2 = \frac{\sum(X - \overline{X}_T)^2}{n_T - 1}$$

The numerator of this formula is called the "total sum of squares," abbreviated SS_T. It's the basis for all the partitioning that will follow. Notice, too, that the formula uses $\bar{X}_T$ as the symbol for the overall mean of all scores (some authors use "GM," for "Grand Mean"), and n_T, the total number of subjects. The denominator of the formula is known as the total degrees of freedom, or df_T. Translating the old variance formula to ANOVA terms, we get:

$$S_T^2 = \frac{\sum (X - \bar{X}_T)^2}{n_T - 1} = \frac{SS_T}{df_T}$$

In ANOVA calculations, this pattern—dividing a sum of squares by an associated df—is repeated again and again. The number that you get when you divide a sum of squares by the appropriate df is called a mean square (MS). So

$$MS_t = \frac{SS_t}{df_t}$$

In a simple ANOVA, the total sum of squares (SS) is broken down into two parts: (1) a *sum of squares within,* SS_W, which reflects the degree of variability within groups but is insensitive to overall differences between the groups, and (2) a *sum of squares between groups,* SS_B, which reflects differences between groups but is not sensitive to variability within groups. The total sum of squares is the sum of the sum of squares within and the sum of squares between: $SS_T = SS_W + SS_B$. The total degrees of freedom can be broken down similarly: $df_T = df_W + df_B$. To find df_W, add up the df's within all the groups: $(n_1 - 1) + (n_2 - 1) + \cdots + (n_{last} - 1)$. And df_B is the number of groups minus 1: $n_G - 1$.[2]

Dividing SS_W by df_W gives us what is known as the mean square within, a measure of the variability within groups:

$$MS_t = \frac{SS_t}{df_t}$$

And dividing SS_B by df_B gives us the mean square between, a measure of variability between groups:

$$MS_B = \frac{SS_B}{df_B}$$

I know you haven't been told how to find SS_W and SS_B yet—that comes next. For now, just look at the logic of the process.

With MS_B we have a measure of variability between the groups, that is, a measure that reflects how different they are from each other. And our MS_W is

[2]Calculating the df's this way allows a neat check of your arithmetic: df_W plus df_B should always be equal to df_T.

a measure of the variability inside the groups, variability that can be attributed to chance or error. Ultimately, of course, we want to know if the between-group differences are significantly greater than chance. So we will compare the two by computing their ratio:

$$F_{\text{obt}} = \frac{\text{MS}_B}{\text{MS}_W}$$

F is the ratio of a mean square between groups to a mean square within groups. (It's named after Sir Roland Fisher, who invented it.) The "obt" subscript means that, as usual, we will compare this obtained value of *F* with some critical value (F_{crit}), which will tell us how likely it is that our F_{obt} could have happened just by chance. The values of F_{crit} are found in Appendix F (I'll show you how to do this later). If F_{obt} is equal to or greater than F_{crit}, then we reject the null hypothesis.

You may have guessed that when comparing three group means the null hypothesis is

$$H_0\colon \mu_1 = \mu_2 = \mu_3$$

The only hard part about ANOVA is learning how to compute SS_W and SS_B. You can do it, though. It's actually easier than some of the other computations we've done.

The scores from our hypothetical study are shown again in Table 9-1. Under the scores you'll see some rows of computations, which I'll explain as we go along. In this computational table, we'll end up with intermediate values, which I've labeled I, II, and III, which we'll use to compute our SS_W and SS_B.

TABLE 9-1

Row		Group 1 22 16 17 18	Group 2 6 10 13 13 8 4	Group 3 8 6 4 5 2	
1	n	4	6	5	$N = \Sigma n = 15$
2	$\Sigma X = T$	73	54	25	$\text{GT} = \Sigma T = 152$
3					$\text{I} = \frac{\text{GT}^2}{N} = \frac{(152)^2}{15}$ **I = 1540.27**
4	ΣX^2	1353	554	145	**II** $= \Sigma\Sigma X^2 =$ **2052**
5	$\frac{T^2}{n}$	$\frac{(73)^2}{4}$ = 1332.25	$\frac{(54)^2}{6}$ = 486	$\frac{(25)^2}{5}$ = 125	$\text{III} = \sum \frac{(T)^2}{n}$ **III = 1943.25**
6	$\bar{X}$	18.28	9	5	

Computation Steps

Row 1. Find N. N stands for the total number of subjects in the entire study. Under each column of scores you will see the number of subjects (n) for each group. In the last column in row 1 you will see N, which is the total number of subjects in the study. You get that, of course, by adding the n's for the columns: $N = n_1 + n_2 + n_3 = 4 + 6 + 5 = 15$.

Row 2. Find GT. GT stands for *grand total* of scores. Start by finding ΣX (which I've called T here) for each column. For example, $\Sigma X = T = 73$ for group 1. Then find GT by adding the totals (T), the sums of the scores, for each column:

$$\text{GT} = 73 + 54 + 25 = 152$$

Row 3. Find **I**. **I** is the first intermediate value we will compute. As the formula indicates, **I** is found by squaring GT (found in row 2) and dividing that by N (found in row 1).

$$\mathbf{I} = 1540.27$$

Row 4. Find **II**. **II** is the total sum of X^2 for the entire study. First, find the ΣX^2 for each column (square each score first, then find the sum of the squares—remember?), and then compute **II** by adding those sums of squares.

$$\mathbf{II} = 1353 + 554 + 145 = 2052$$

Row 5. Find **III**. To find **III**, square T for each column and divide it by n for that column; then find the sum of those values.

$$\mathbf{III} = 1332.25 + 486 + 125 = 1943.25$$

Row 6. This is sort of an "extra" that I added because you're going to need the group means in a minute or two.

Now use the intermediate values to compute the sums of squares with the following formulas:

$$\text{SS}_B = \mathbf{III} - \mathbf{I} = 1943.25 - 1540.27 = 402.98$$

$$\text{SS}_W = \mathbf{II} - \mathbf{III} = 2052 - 1943.25 = 108.75$$

We will also need to know the degrees of freedom for both between and within groups. To use the next formulas, you need to know that k is the number of groups. In our example, $k = 3$. You will also need to recall that N is the total number of subjects in the study:

$$\text{df}_B = k - 1 = 3 - 1 = 2$$

$$\text{df}_W = N - k = 15 - 3 = 12$$

Now we can fill in what is known as an ANOVA summary table (see Table 9-2).

TABLE 9-2 One-way ANOVA Summary Table

Source of Variation	Degrees of Freedom (df)	Sum of Squares	Mean Squares	F
Between (B)	$k - 1 = 3 - 1 = 2$	$SS_B = \textbf{III} - \textbf{I}$ $= 402.98$	$MS_B = \frac{SS_B}{df_B}$ $= \frac{402.98}{2}$ $= 201.49$	$F_{obt} = \frac{MS_B}{MS_W}$ $= 22.24^{**}$
Within (W)	$N - k = 15 - 3$ $= 12$	$SS_W = \textbf{II} - \textbf{III}$ $= 108.75$	$MS_W = \frac{SS_W}{df_W}$ $= \frac{108.75}{12}$ $= 9.06$	
Total	$N - 1 = 14$	$SS_T = 511.73$		

As was true for the t test, to find out whether your F_{obt} is statistically significant, you will need to compare it with F_{crit}. You find F_{crit} in Appendix F. Go across the table until you come to the column headed by the df_B for the F_{obt} you are interested in (in our example, there is only one F_{obt} to worry about, and $df_B = 2$). Follow down that column until you are directly across from your df_W (in our example, $df_W = 12$). Appendix F includes critical values of F for both the .05 and the .01 levels of significance; the .01 level is in boldface type. As you can see, for $df_B = 2$ and $df_W = 12$, F_{crit} at the .05 level of significance is 3.88, and at the .01 level F_{crit} is 6.93. Since your $F_{obt} = 22.24$ is obviously larger than F_{crit} for either the .05 or the .01 levels of significance, you reject the null hypothesis (that there is no difference among means) and conclude that at least one of the means is significantly different from at least one of the others. Conventionally, a value of F that exceeds F_{crit} for the .05 level is followed by a single asterisk; if it exceeds F_{crit} for the .01 level it gets two asterisks (see Table 9-2).

Looking at the means of the three groups in our example, it appears that the mean of group 1, the counseled group ($\overline{X} = 18.25$) is significantly different from that of the group 3, the control group ($\overline{X} = 5$). We might wonder if it is also significantly different from group 2, the exercise/diet group ($\overline{X} = 9$), or if the exercise/diet group is significantly different from the control group. To answer these types of questions, you will need to learn how to do what is known as post hoc analysis. One method of doing this kind of analysis is found in the next section.

POST HOC ANALYSIS

The procedure in this section is designed to be used after an ANOVA has been done and the null hypothesis of no difference among means has been rejected. For example, consider a study of five different teaching methods. Five groups of students were taught a unit, each group being exposed to a different teaching method, and then the groups of students were tested for how much they had learned. Even if an ANOVA were to show the differences among the five means to be statistically significant, we still would not know which of the pairs of means were significantly different: Is $\overline{X}_1$ significantly different from $\overline{X}_2$? How about the difference between $\overline{X}_2$ and $\overline{X}_3$? If we looked at each possible combination, we would have $5(51)/2 = 10$ pairs of means to analyze. You will recall from the discussion at the beginning of this chapter that it is not good practice to analyze differences among pairs of means with the *t* test because of the increased probability of a Type I error; the same criticism can be leveled at any large set of independent comparisons.

Many procedures have been developed to do what is called post hoc analysis (tests used after an F_{obt} has been found to be statistically significant in an ANOVA). This book will present only one of these methods, the Sheffé, which can be used for groups of equal or unequal *n*'s.

THE SHEFFÉ METHOD OF POST HOC ANALYSIS

The statistic you will compute in the Sheffé method is designated as *C*. A value of *C* is computed for any pair of means that you want to compare; unlike the *t* test, *C* is designed to allow multiple comparisons without affecting the likelihood of a Type I error. Moreover, if all the groups are the same size, you don't have to compute *C* for every single pair of means; once a significant *C* has been found for a given pair, you can assume that any other pair that is at least this far apart will also be significantly different. (With a *t* test, this is not necessarily true; nor is it always true for *C* when the group *n*'s are unequal.)

As is the usual procedure, you will compare your C_{obt} with a C_{crit}; if C_{obt} is equal to or greater than C_{crit}, you will reject the null hypothesis for that pair of means. In the Sheffé test you don't look up C_{crit} in a table; I'll show you how to compute it for yourself. First, though, we'll deal with C_{obt}.

$$C_{obt} = \frac{\overline{X}_1 - \overline{X}_2}{\sqrt{MS_W\left(\frac{1}{n_1} + \frac{1}{n_2}\right)}}$$

where $\overline{X}_1$ and $\overline{X}_2$ are the means of two groups being compared
n_1 and n_2 are the *n*'s of those two groups
MS_W is the within-group mean square from your ANOVA

Now let's go back to our three groups, counseled (group 1), exercise/diet (group 2), and control (group 3). Since we were able to reject H_0, we know that at least one group is significantly different from one other group; but we don't know which groups they are. And there may be more than one significant difference; we need to check that out. We'll start with the first pair, group 1 versus group 2.

$$C_{obt} = \frac{\overline{X}_1 - \overline{X}_2}{\sqrt{MS_W\left(\frac{1}{n_i} + \frac{1}{n_j}\right)}} = \frac{18.28 - 9}{\sqrt{9.06\left(\frac{1}{4} + \frac{1}{6}\right)}} = \frac{9.28}{1.94} = 4.77$$

For group 1 versus group 3,

$$\frac{18.28 - 5}{\sqrt{9.06\left(\frac{1}{4} + \frac{1}{5}\right)}} = \frac{13.28}{2.02} = 6.57$$

And for group 2 versus group 3,

$$\frac{9 - 5}{\sqrt{9.06\left(\frac{1}{6} + \frac{1}{5}\right)}} = \frac{4}{1.82} = 2.2$$

(Notice that, although it's still tedious to look at each possible pair, it's a lot less work than doing multiple t's!)

Now we are ready to compute C_{crit}. At the .05 level of significance ($\alpha = .05$),

$$C_{crit} = \sqrt{(k-1)(F_{crit})} = \sqrt{(3-1)(3.88)} = 2.79 \text{ where } k = \text{ the number of groups}$$

When $\alpha = .01$ and $k = 3$,

$$C_{crit} = \sqrt{(k-1)(F_{crit})} = \sqrt{(3-1)(6.93)} = 3.72$$

Since it doesn't matter which group mean is subtracted from which in each computation of C_{obt}, the sign of the value you get doesn't matter either. Just treat C_{obt} as if it were positive. As you can see by comparing C_{crit} with C_{obt}, for both the .05 and the .01 levels of significance, the mean of the counseled group (group 1) is significantly larger than either the exercise/diet or the control groups. However, the mean of the exercise/diet group is not significantly greater than that of the control group.

PROBLEMS

1. A researcher is interested in differences between blondes, brunettes, and redheads in terms of introversion/extroversion. She selects random samples from a col-

lege campus, gives each subject a test of social introversion, and comes up with the following:

Blondes	Brunettes	Redheads
5	3	2
10	5	1
6	2	7
2	4	2
5	3	2
3	5	3

Use a simple ANOVA to test for differences among the groups.

2. A (hypothetical!) study of eating patterns among people in different occupations yielded the following:

	Bus Drivers (n = 10)	College Professors (n = 10)	U.S. Presidents (n = 4)
Mean junk food score	12	17	58.3

Source	df	Sums of Squares	Mean Squares	F
Between	2	6505	3252.5	6.1**
Within	21	11197	533.2	

a. What do you conclude?
b. Perform the appropriate post hoc analysis.

Answers

1. (I've given you some intermediate steps here, in case you got confused.)

	Blondes	Brunettes	Redheads	
n	6	6	6	18
$\Sigma X = T$	31	22	17	GT = 70
I				$\frac{GT^2}{N} = \frac{70^2}{18} = 272.22$
II	$\Sigma X^2 = 199$	$\Sigma X^2 = 88$	$\Sigma X^2 = 71$	$\Sigma\Sigma X^2 = 358$
III	$\frac{31^2}{6} = 160.17$	$\frac{22^2}{6} = 80.67$	$\frac{17^2}{6} = 48.17$	289.01
$\overline{X}$	5.17	3.67	2.83	

Source	df	Sums of Squares	Mean Squares	F
Between	2	16.79	8.4	1.8
Within	15	68.99	4.6	

Since F_{crit} for $\alpha = .05$, with 2 and 15 df, is 3.68, the obtained value of F does not reach the critical value and we cannot reject H_0.

2. The null hypothesis can be rejected at the .01 level; the differences among the groups are significant ($p < .01$). Scheffé's test, using C_{crit} of 2.63 (for $\alpha = .05$) indicates no significant differences between bus drivers and professors ($C < 1$), but significant differences between college professors and U.S. presidents ($C = 3.03$).

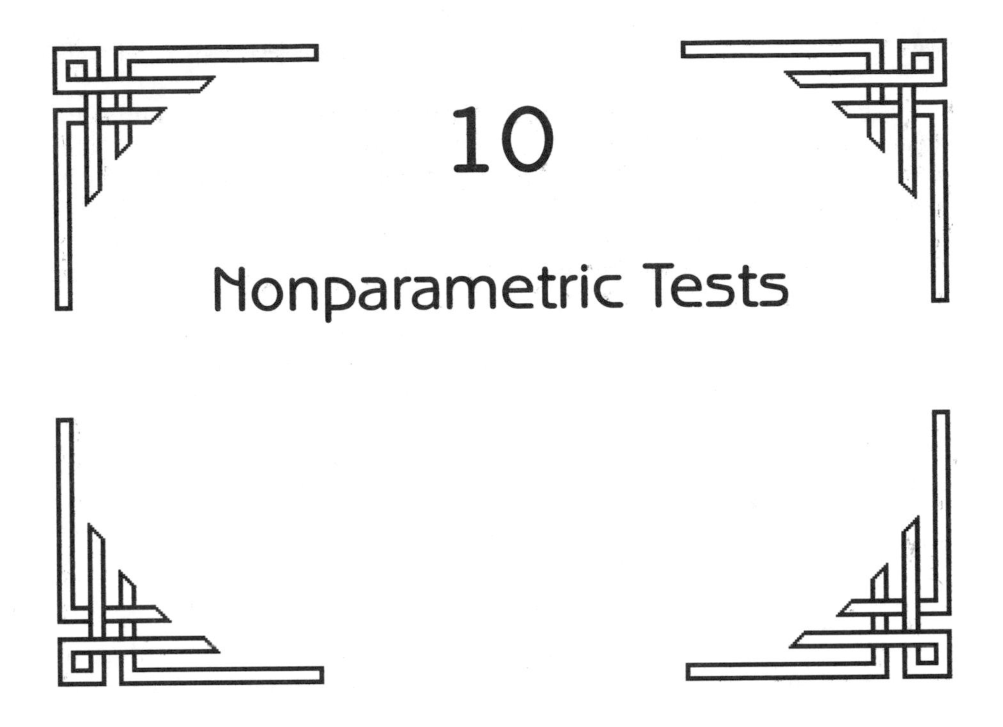

10

Nonparametric Tests

"Nonparametric"—oh, no! Is this something I should recognize from what has gone before? If we're going to look at *non*parametric things, what in the world are parametric ones? Relax—all these questions will get answered. You should know, though, first of all, that the things we will be doing in this chapter will be easier, not harder, than what you have already done.

All the statistical procedures that you have learned so far are *parametric* procedures. The idea of parametric involves a number of things, but the most important one is that the data used in parametric tests or techniques must be scores or measurements of some sort. When comparing groups of people by means of a parametric test, we measure all the people and then use the test to determine whether the groups' measurements are significantly different from what could be expected just by chance.

But what if we are dealing with a situation in which we don't measure or test our subjects? That's not as unusual as you may imagine. Here's an example: a researcher wants to know if men and women college students differ in terms of their use of the college Counseling Center. She randomly selects 100 students and asks them if they have ever used the center. Here are her findings:

	Men	Women	Totals
Have used the center	12	16	28
Have not used the center	48	24	72
Totals	60	40	100

How should this researcher analyze her data? There are no scores, no means or variances to calculate. What she has are four *categories:* men and women who either used or did not use the center. Each person in the sample can be assigned to one, and only one, of these categories. What the researcher wants to know is whether the distribution she observed is significantly different from what she might expect, by chance, if the total populations of men and women didn't differ.

Her question is quite easy to answer, using a procedure called the chi-square test. Chi square (symbolized by the Greek letter chi, squared: χ^2) is a nonparametric test. It doesn't require that its data meet the requirements of the parametrics, and it most particularly doesn't require that the data be in the form of scores or measurements. Instead, it was specifically designed to test hypotheses about category data.

The null hypothesis for our example is that men and women don't differ significantly in their use of the Counseling Center:

$$H_0\text{: Men's use of the center} = \text{Women's use of the center}$$

To test it using chi square, we first need to figure out the distribution that we would most often get just by chance if H_0 were true. Because there are not an equal number of men and women in the total sample, we can't say that H_0 predicts that half of the center's clients will be male and half female. Rather, since 60% of the sample are male, H_0 predicts that 60% of the people who used the counseling center ought also be male and 40% female. Similarly, 60% of the people who didn't use it should be male and 40% female, if H_0 is true. We call these numbers the ***expected*** frequencies, or f_e. The observed frequencies, not surprisingly, are designated f_o.

$$f_e \text{ for men who have used the center is } .6(28) = 16.8$$

$$f_e \text{ for men who have not used the center is } .6(72) = 43.2$$

$$f_e \text{ for women who have used the center is } .4(28) = 11.2$$

$$f_e \text{ for women who have not used the center is } .4(72) = 28.8$$

Notice that the f_e's yield the same row and column totals as the f_o's, providing you with a good way to check your math!

	Men	Women	Totals
Have used the center	$f_o = 12$ / $f_e = 16.8$	$f_o = 16$ / $f_e = 11.2$	28
Have not used the center	$f_o = 48$ / $f_e = 43.2$	$f_o = 24$ / $f_e = 28.8$	72
Totals	60	40	100

The formula for those f_e's is

$$f_e \text{ of cell } i, j = \frac{T_{\text{column } j}}{\text{GT}}\left(T_{\text{row } i}\right)$$

where $T_{\text{column } j}$ is the total number of people in column j
$T_{\text{row } i}$ is the total number of people in row $_i$
GT is the grand total, the number of people in the whole study

And do notice that this formula will work no matter how many rows and columns you have: it would be perfectly fine to use it with data collected on blonds, brunettes, and redheads who are majoring in either physics, chemistry, psychology, P.E., or English literature. Come to think of it, that would be an interesting problem. I think I'll set it up for you at the end of the chapter.

Once the f_e's have been calculated, it is a simple matter to find the value of χ^2:

$$\chi^2 = \Sigma \frac{\left(f_o - f_e\right)^2}{f_e}$$

1. Within each cell, find the value of $(f_o - f_e)$.

$12 - 16.8 = -4.8$

$48 - 43.2 = 4.8$

$16 - 11.2 = 4.8$

$24 - 28.8 = -4.8$

2. For each of the differences, square and divide by f_e.

$$\frac{-4.8^2}{16.8} = 1.37$$

$$\frac{-4.8^2}{43.2} = .53$$

$$\frac{-4.8^2}{11.2} = 2.06$$

$$\frac{-4.8^2}{28.8} = .80$$

3. Add up your answers.

$$\chi^2 = 1.37 + .53 + 2.06 + .80 = 4.76$$

As has been the case with all our statistics, we must compare the calculated value of chi square with a critical value. To do so, enter Appendix G with df $= (r - 1)(c - 1)$: the number of rows minus 1 times the number of columns minus 1. In our example, $r = 2$ and $c = 2$, so df $= 1 \times 1 = 1$. You can see from the table that, for $\alpha = .05$,

$$\chi^2_{\text{crit}} = 3.84$$

Since our observed value of χ^2 (4.76) is greater than the critical value, we reject the null hypothesis. It is reasonable to conclude that men and women do differ in their use of the Counseling Center. Note, though, that this result does not imply any sort of causal relationship: we cannot say that gender "causes" the differences in Counseling Center behavior, any more than we could say that use of the Counseling Center "causes" people to be either male or female.

Here's another example: suppose that you wished to evaluate the effectiveness of three different methods of counseling, methods A, B, and C. Also suppose that the best measure of outcome that you could get was a categorization of "improved" or "not improved." Sixty clients were randomly assigned to and counseled by each method, with results as follows:

	Method A	Method B	Method C	Totals
Improved	10	18	12	40
Not improved	10	2	8	20
Totals	20	20	20	60

Begin by calculating the values of f_e, using the same formula as before. I'll give you the first one:

$$f_e \text{ of cell } 1,\ 1 = \frac{T_{\text{column }1}}{\text{GT}} T_{\text{row }1} = \frac{20}{60} 40 = 13.3$$

Do the rest of them yourself before looking at my table.

	Method A	Method B	Method C	Totals
Improved	$f_o = 10$ / $f_e = 13.3$	$f_o = 18$ / $f_e = 13.3$	$f_o = 12$ / $f_e = 13.3$	40
Not improved	$f_o = 10$ / $f_e = 6.7$	$f_o = 2$ / $f_e = 6.7$	$f_o = 8$ / $f_e = 6.7$	20
Totals	20	20	20	60

Now, compute chi square:

$$\chi^2 = \Sigma \frac{(f_o - f_e)^2}{f_e}$$

$$= .82 + 1.67 + .12 + 1.63 + 3.30 + .25$$

$$= 7.78$$

In this example we have 2 rows and 3 columns, so

$$df = (2 - 1)(3 - 1) = 2$$

Appendix G shows $\chi^2_{crit} = 5.99$ for an α of .05; and 9.21 for an α of .01. You can reject the null hypothesis at the .05 level and conclude that there are significant differences in the success rates among the three counseling methods.

Generally, studies utilizing chi square will require a relatively large number of subjects. Many statisticians recommend a certain minimum ***expected*** frequency (f_e) per cell. A very conservative rule of thumb is that f_e must always be equal to or greater than 5.

THE MANN–WHITNEY *U* TEST

Not infrequently, researchers find themselves collecting data in the form of ranks rather than scores: we can say whether a particular person did better/more/longer than another, but not ***how much*** better/more/longer. The members of the sample can be put in rank order, from top to bottom, but can't be given actual score values. Parametric tests should not be used with ranked information like this; parametrics require that each score be a direct measurement of the subject, rather than just an indication of where that subject stands relative to the others. To determine whether two groups of ranked data are significantly different from each other, we need to use a nonparametric technique. One of the most commonly used of these is the Mann–Whitney *U* test.

I'll use an example to show you how the *U* test works. A teacher was asked to rank his second grade students in terms of their social skills, and he provided the following information:

Judith	1	(highest skill level)
Sam	2	
Kenny	3	
Bruce	4	
Brian	5	
Marilyn	6	
Loice	7	
Elaine	8	
Holly	9	
Hope	10	
Grant	11	
Bobby	12	
Toni	13	
Meredith	14	
Sam	15	(lowest skill level)

Our researcher is interested in possible social skill differences between children who change schools versus children who stay in the same school. Of this class, seven children had transferred in from other school districts: Bruce, Elaine, Holly, Grant, Bobby, Meredith, and Sam.

The null hypothesis here is that the rankings of the transferred children are not significantly different from those of the nontransferred children, H_0: $R_{trans} = R_{nontrans}$. To test this hypothesis, we first divide the total class into transferred and nontransferred groups.

Nontransferred		Transferred	
Name	Rank	Name	Rank
Judith	1	Bruce	4
Sam	2	Elaine	8
Kenny	3	Holly	9
Brian	5	Grant	11
Marilyn	6	Bobby	12
Loice	7	Meredith	14
Hope	10	Sam	15
Toni	13		
Sum of ranks	47		73

Notice that we kept the ranks that were given when the children were in a single sample. If we had started out with two separate groups, we would have to combine them and assign ranks for the combined sample. Then we would reseparate them into their original groups, keeping track of each child's combined-group rank.

Next, choose one of the two groups (it doesn't matter which one) and add up their ranks; then plug that sum into the formula for U:

$$U_1 = N_1 N_2 + \frac{N_1(N_1+1)}{2} - T_1$$

where N_1 and N_2 are the numbers of people in each group
T_1 is the total of the ranks for the group you've chosen

Just for practice, let's compute the value of U for both of our groups:

$$U_1 = (8)(7) + \frac{(8)(9)}{2} - 47 = 45$$

$$U_2 = (8)(7) + \frac{(7)(8)}{2} - 73 = 11$$

You will probably not be surprised to learn that we are going to compare our obtained value of U with a U_{crit} that we'll look up in a table. But which of our U's do we use (U's use? use U's? forget it . . .)? We use the smaller of the two. If we don't want to compute both of them directly, we can find the value of the smaller by

$$U_2 = N_1 N_2 - U_1$$
$$= (8)(7) - 45 = 56 - 45 = 11$$

And this also provides us with a handy check of our arithmetic.

We use the smaller value of U because with the Mann–Whitney test, unlike the other statistical tests you've learned, a *smaller* obtained value is more likely to be significant. You use the table in Appendix H to find out if your obtained value of U is small enough to indicate that the two groups of children are more different from each other than would be expected by chance alone.

There are a number of different ways to organize a table of Mann–Whitney U critical values, none of which is completely satisfactory; I've chosen the one that seems easiest to use. But it's different from the other tables you've seen in this book, in three important ways:

1. It's really a series of tables, one for each possible value of N_2 (the size of the larger group). So you need to be sure you're entering the right subtable.
2. The body of the table contains the exact probability of obtaining a given U value just by chance.
3. It contains one-tailed values; that is, the values are appropriate only when you have been able to predict ahead of time which group will be ranked higher. If you haven't made a one-tailed prediction, you will need to double the probability values in the body of the table.

For our example, we'll want the subtable for $N_2 = 8$. Read across the top to find the column for the size of the other group ($N_1 = 7$); when you've found it, follow down to the row for the value of U that we calculated, $U = 11$. The probability of obtaining a value of U this small just by chance is only .027. Can we reject H_0 at the .05 level? We could if we were doing a one tailed test. But we didn't predict ahead of time which group of children would do better, and so we must use a two-tailed test. Twice .027 is .054; we cannot reject the null hypothesis at the .05 level.

But what if one of our groups has an N of more than 8? With larger groups, U converts to z, and we look in the normal curve table for significance levels. Simply plug your numbers into the formula

$$z = \frac{U_1 - \frac{(N_1)(N_2)}{2}}{\sigma_U}$$

where

$$\sigma_U = \sqrt{\frac{(N_1)(N_2)(N_1 + N_2 + 1)}{12}}$$

This is another one that looks hard, but isn't. Imagine, for instance that you have two groups with N of 7 and 9, respectively. You've calculated that the value of the smaller $U = 15$.

$$\sigma_U = \sqrt{\frac{(7)(9)(17)}{12}} = \sqrt{\frac{1071}{12}} = \sqrt{89.25} = 9.4$$

and

$$z = \frac{15 - \left(\frac{7 \cdot 9}{2}\right)}{9.4} = \frac{15 - 31.5}{9.4} = -1.7$$

Looking in Appendix B, we find that a z of -1.7 or less, *or of 1.7 or more* (remember, since we didn't specify a predicted direction of difference, we must do a two-tailed test), will occur only .0446 + .0446 or .0892 of the time just by chance. Since .0892 > .05, we cannot reject H_0.

THE SPEARMAN CORRELATION FOR RANKED DATA

You have now been introduced to nonparametric techniques that allow you to test for differences between groups when your data are in the form of categories (chi square) and ranks (U test). The last nonparametric technique we will discuss is also used with ranked data, but it allows us to look directly at the degree to which two variables are related. It's called the Spearman correlation for ranked data, symbolized by r_S.

Remember when, a long time ago, we talked about correlations and learned to compute the value of r? The correlation coefficient we used, r, gave us a measure of the amount of relationship between two variables. But r, like all the parametric statistics, is assumed to be based on measurements that are independent of each other. It follows that r would be inappropriate to use with ranked data, since ranks are neither measurements nor independent. So what do you do when the variables you want to know about come in the form of rankings? Enter the Spearman correlation for ranked data, r_S.

Like r, r_S can range between -1.00 and $+1.00$. Also like r, values of r_S near zero indicate little relationship between the two variables; values close to -1.00 or 1.00 indicate a near-perfect relationship. And finally, like r, it is possible to determine whether an obtained r_S is large enough to reject the null hypothesis (that is, that the variables are not related).

Our example: at the beginning of their graduate training, a group of counseling students watched video tapes of 10 counseling sessions; their task was to discuss the sessions among themselves and then rank them according to how helpful they judged the counselor's behavior to be. At the end of the first year of their program, the group was called together again and asked to redo their ratings. The researcher wanted to know how closely the ratings of the untrained group resembled its ratings after a year of training. Here is what she found:

Counseling Session	Rating 1	Rating 2	D	D^2
A	1	2	−1	1
B	2	1	1	1
C	3	8	−5	25
D	4	3	1	1
E	5	6	−1	1
F	6	4	2	4
G	7	7	0	0
H	8	5	3	9
I	9	10	−1	1
J	10	9	1	1

Ignore the two right-hand columns for now. Notice that we started out with the counseling sessions arranged in rank order, from lowest ranking to highest. At the end of the year, the rankings have shifted a bit; most noticeably, the session that was originally ranked third is now number 8. Just how similar are the two sets of rankings? We can compute r_S using the following formula:

$$r_s = 1 - \frac{6\sum D^2}{n(n^2 - 1)}$$

D stands for the difference between each pair of ranks, and D^2 is the square of those differences. n is the number of pairs of rankings. For our data,

$$r_s = 1 - \frac{6 \cdot 44}{10(100 - 1)} = 1 - \frac{264}{990} = 1 - .27 = .73$$

Although r_S is not an exact analog to r, we can still evaluate it in roughly the same way. A correlation of .73 seems quite respectable, whether it is an r or an r_S. But appearances can sometimes be deceiving. The "respectability" of a value of r *or* r_S depends on how many pairs of values have been considered, since smaller numbers of measurements can often produce high correlations just by chance. Appendix I gives the critical values for r_S. In this table, n again refers to the number of pairs. With $n = 10$ and $\alpha = .05$, any value of r_S greater than .564 is significant and allows us to reject the null hypothesis. We concluded that our original impression was correct, that the two sets of ratings are more closely related than would be expected by chance alone.

A BIT MORE

This chapter has been a sort of grab bag, a description of three tools for dealing with data that can't be handled by parametric techniques. There are lots more—three of them is just a teaser to whet your appetite! It's safe to say that a nonparametric exists, or can be cobbled to fit, virtually any research design.

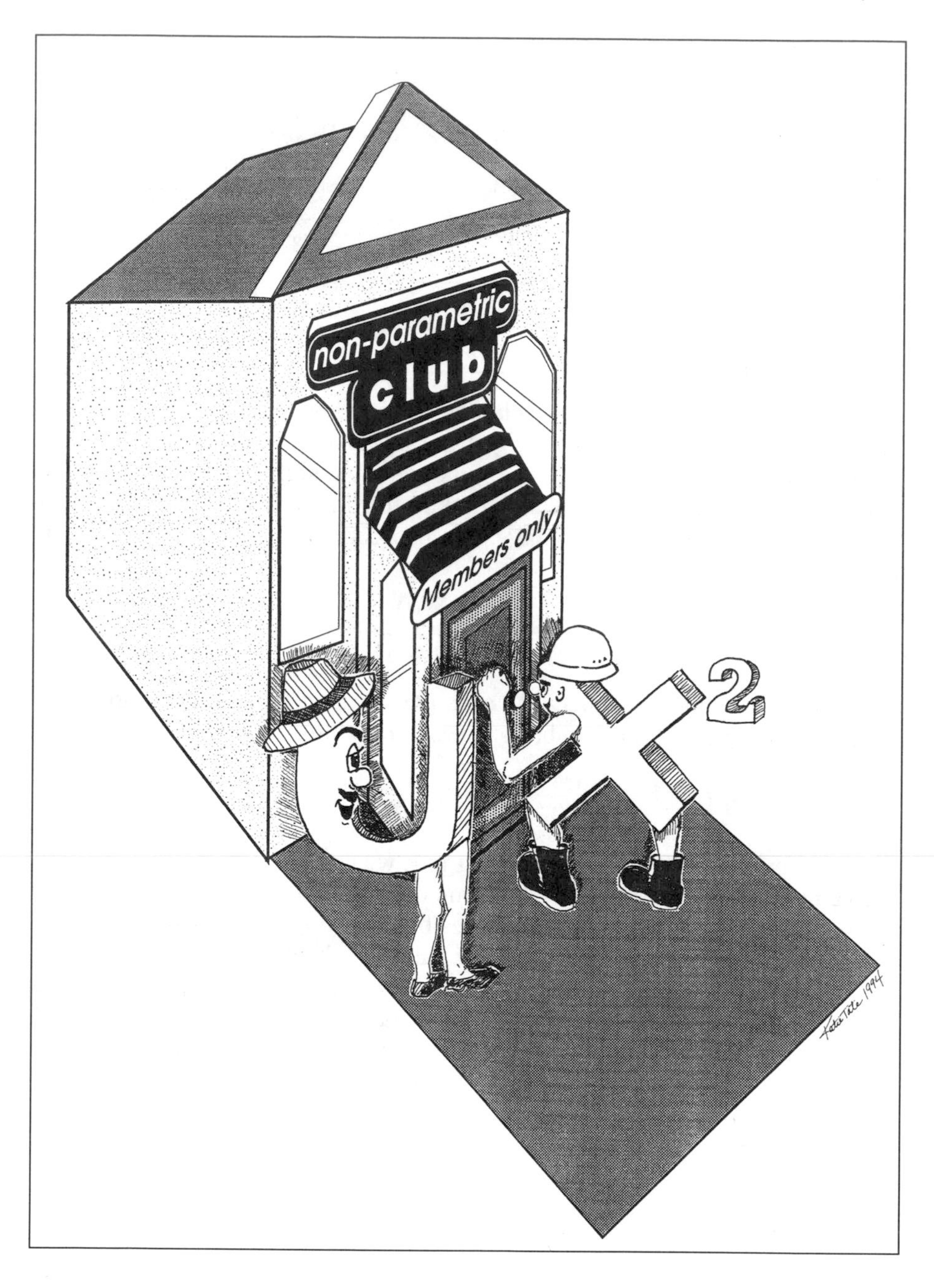
non-parametric
club
Members only
2

Additionally, nonparametrics tend to have a relatively simple structure, to be based on less complex sets of principles and assumptions than their parametric brethren. For this reason, if you can't find a ready-made nonparametric test that fits your situation, a competent statistician can probably design one for you without much trouble.

At this point you may be asking, "So why don't we always use nonparametrics? Why bother with parametric tests at all?" The answer is simple: parametric tests are more powerful than nonparametrics. When the data allow it, we prefer to use parametrics because they are less likely to cause a Type II error. Another way of saying this is that parametric tests let us claim significance more easily; they are less likely to miss significant relationships when such relationships are present. However, using a parametric test with data that don't fit can cause even more problems—can result in Type I error—and we can't even estimate the probability of Type I error under these circumstances. That's when we need the nonparametrics. Not only are they, by and large, easier to compute, but they fill a much needed spot in our statistical repertoire.

PROBLEMS

1. To explore differences in seating preferences between male and female college students, you count who is sitting where in a large introductory English class.

	Males	Females
Front part of the room	12	26
Back part of the room	35	20

Do a χ^2 test to determine if the differences are significant; use $\alpha = .05$.

2. Here's the one I promised you back on p. 101. A researcher randomly selects a sample of college students majoring in physics, chemistry, psychology, P.E., or English lit. He notes each student's major subject, and also whether that student is a blond, a brunette, or a redhead. The data are as follows:

	Physics	Chem	Psych	P.E.	English	Total
Blond	8	10	20	4	10	52
Brunette	10	8	30	3	10	61
Redhead	5	3	10	14	10	42
	23	21	60	21	30	155

Can the researcher conclude that the relationship between major and hair color is greater than would be expected by chance alone?

3. You randomly selected 10 men and 10 women from a large college class, mixed up their names, and asked the professor to put them in order in terms of attentive-

ness. After some grumbling, the professor agreed; the next week he arranged the names as follows (most attentive first):

1. Ann	11. Brad
2. Sue	12. Jane
3. George	13. Claire
4. Mary	14. Harry
5. Sam	15. Lucy
6. Karen	16. Bruce
7. Darren	17. Neil
8. Christopher	18. Elaine
9. Emily	19. Rudy
10. Marie	20. Alex

Why must you use a nonparametric test to determine whether there is a significant difference between the men and the women in terms of attentiveness? What test will you use? (You know what's coming . . .) Do it!

4. You learn that these same 20 students are also taking a beginning math class. Is there a relationship between attentiveness level in the two classes? You get the math professor to do the same kind of ranking as was done in the English class, and the data look like this:

1. Ann	11. Jane
2. Alex	12. Karen
3. Brad	13. Lucy
4. Chris	14. Claire
5. Marie	15. Mary
6. Darren	16. Emily
7. Bruce	17. Harry
8. George	18. Neil
9. Elaine	19. Sam
10. Sue	20. Rudy

What is the correlation between the two sets of rankings? Is the relationship between behavior in the two classes significant?

Answers

1. $\chi^2 = 9.2$. χ^2_{crit} with $\alpha = .05$ and 1 df is 3.84; the relationship between seating and gender is significant.
2. $\chi^2 = 24.5$. χ^2_{crit} with $\alpha = .01$ and 8 df is 20.09; the relationship between hair color and major is significant.
3. A nonparametric test must be used because the data are rankings, not independent measurements. $U_1 = 35$; $N_1 = N_2 = 10$; $z = 1.14$; H_0 cannot be rejected.[1]
4. $r_S = .17$; the relationship is not significantly different from what might be expected to occur by chance alone.

[1]We had to calculate z because the groups were too large for the U table in Appendix H. And if you managed to claw your way through this whole line of numbers and symbols and could actually figure out what they meant—wow! You really have come a long way, Baby!

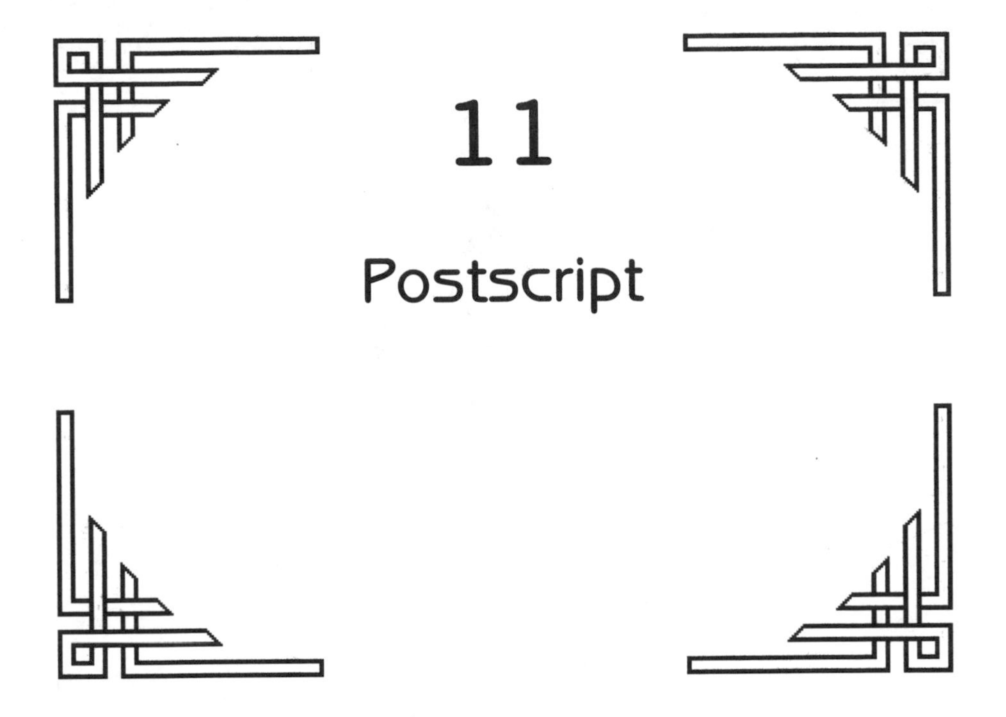

11 Postscript

Well, you did it! You got all the way through this book! Whether you realize it or not, this means that you have covered a great deal of material, learned (probably) quite a bit more than you think you did, and are now able to do—or at least understand—most of the statistics that you will need for a large percentage of the research that you may become involved with. No small accomplishment!

In case you are inclined to discount what you've done, let's review it: this will not only give you further ammunition for self-congratulation, but will also help to consolidate all the information you've been taking in.

Statistics, as a field of study, can be divided into two (not so equal) parts, descriptive and inferential. You've been introduced to both. First, the descriptive.

You've learned how to describe sets of data in terms of graphs (histograms, frequency polygons, cumulative frequency polygons), of middleness (mean, median, mode), and of spread (range, variance, standard deviation). You've also learned that a distribution—and even that word was probably somewhat unfamiliar when you began all this—can be symmetrical or skewed, and you've learned what happens to the middleness measures when you skew a distribution.

You've also learned a lot about how two distributions—two sets of data—can be related. You learned how to compute a correlation coefficient and what a correlation coefficient means. You learned what a scatter plot is

and how the general shape of a scatter plot relates to the value of *r*. You learned how to use something called a regression equation to predict a score on one variable, based on an individual's performance on a related variable, and to use something called the standard error of the estimate to tell you how much "play" there is in that prediction. You learned how to compute a correlation coefficient on data comprised of ranks instead of measurements. You even learned how to calculate the relationship among more than two variables and to partial out the influence of one variable in order to get a clearer idea about the relationship between two others.

And then along came inferential statistics: using a set of observable information to make inferences about larger groups that can't be observed. You started this section by absorbing a lot of general ideas and principles. You learned about sampling, why it's important that a sample be unbiased, and how to use random sampling techniques to get an unbiased sample. You learned what a null hypothesis is, why we need to use a null hypothesis, and what kinds of error are associated with mistakenly rejecting or failing to reject the null hypothesis. You learned that it's a great big No–No to talk about "accepting" the null hypothesis and why! You learned what it means for a result to be statistically significant, and you got acquainted with a friendly Greek named α.

Then you moved into actual statistical tests, starting with the *t* test. Using the *t* test, you now know how to decide whether two groups are significantly different from each other, and you know that correlated or matched groups have to be treated differently from independent groups. You also know what a one-tailed test and a two-tailed test are and when it's appropriate to use each kind.

As if looking at two groups weren't enough, you moved right in to explore comparisons among three or more groups. You learned about the workhorse of social science statistics, the analysis of variance. (Although we didn't have time to talk about it, the procedures you learned in dealing with the one-way ANOVA are the basis for much more complicated ANOVA designs, which you are now well prepared to master). And you learned how to do a post hoc test, to look at those three or more groups in even more detail.

Finally, you rummaged around in a bag of tricks designed to allow you to work with data that don't fit the rules for the other techniques that you've learned. You learned a fine, important sounding new word: *nonparametric*. You learned nonparametric tests that let you examine frequencies and ranked data.

You really have done a lot!

And we haven't even mentioned the single most important, and impressive, thing you've done. Imagine your reaction just a few months ago if someone had handed you the first paragraphs of this chapter and said, "Read this—this is what you will know at the end of this term." The very fact of your being able to think about statistics now without feeling frightened or overwhelmed or nauseous is much more significant than the facts and techniques that you've learned. Because your changed attitude means that you're able to actually use all this stuff, rather than just being intimidated by it. If you can't remember some statistical something now, you can go look it up, instead of giving up. If you can't find out where to look it up, you can ask somebody and have a reasonable expectation of understanding their answer.

Moreover, you're ready to move on to the next level: you've poured yourself a good, solid foundation that you can build on, just about as high as you want to go. There is more to statistics than we've been able to cover, of course. There are all the mathematical implications and "pre-plications" (well, what else do you call something that comes before and influences the thing you're interested in?) of the techniques you've learned. There are the fascinating nooks and crannies of those techniques—the sophisticated rules about when to use them, the exceptions to the rules, the suggestions for what to do instead. There are the extensions: applying the basic principles of correlation or regression or ANOVA to larger and more complex designs. And then there are the brand new techniques, things like analysis of covariance, and factor analysis, and multiple regression, and lots and lots of clever nonparametric tricks. Why, you might even learn to *enjoy* this stuff!

But whether you learn to enjoy it or not, whether you go on to more advanced work or just stay with what you now know, whether you actively use your statistics or simply become an informed consumer—whatever you do with it—nobody can change or take away the fact that you did learn it, and you did survive. And it wasn't as bad as you thought it would be—truly, now, was it?

APPENDIX A

Basic Math Review

Being terrified of math didn't just happen to you overnight. Chances are that you have been having bad experiences with math for many years. Most people who have these sorts of bad experiences have not learned some of the basic rules for working with numbers. Because they don't know the rules, the problems don't make sense—sort of like trying to play baseball without knowing the difference between a ball and a strike, or what it means to steal a base.

When the problems don't make sense, but everyone else seems to understand them, we are likely to decide that there's something wrong with us. We'll just never be able to do it and, besides, we hate math anyhow. So we tune out, turn off—and the bad situation gets worse.

This book is designed to help you get past that kind of head space. The appendix you're starting now will give you a chance to review the rules that you need in order to play with numbers and come up with the same answers as everyone else. Some of the material will be very familiar to you; other parts may seem completely new. Let me make a few suggestions about how to use this appendix:

1. If, after a couple of pages or so, you're completely bored and have found nothing you don't already know, just skim through the rest and get on with Chapter 1.
2. If it seems familiar, but you're still a little shaky, go to the problems at the end of each section and work them through. That will tell you which parts you need to spend more time on.
3. If much of it feels new to you—take your time with it! Most of us "word people" absorb numerical information and ideas quite slowly and need lots of practice before it really sinks in and becomes part of our way of thinking. Give it a chance. Learning the rules now will allow you to understand the rest of the book in a way that will probably surprise you.

I've divided the basic rules into six sections: (1) positive and negative numbers, (2) fractions and percents, (3) roots and exponents, (4) order of computation, (5) summation, and (6) equations.

POSITIVE AND NEGATIVE NUMBERS

In a perfect and orderly world, all numbers would be positive (they'd be whole, too—no fractions or decimals). But the world isn't perfect, and negative numbers have to be dealt with. Actually, they're not so bad; you just have to show them who's boss.

If you're a visually oriented person, it may help to think of numbers as standing for locations on a straight line, with zero as your starting point. The number 2 is 2 steps out from zero; add 4 more steps and you're 6 steps out, and so on. The negative numbers are just steps in the opposite direction. If I'm 6 steps away from zero, and I add –4 to that, I take 4 steps back; now I'm at 2.

If you're not visually oriented, the last paragraph may confuse you; if so, just ignore it and follow the rules I'm about to give you.

Rule 1. If a number or an expression is written without a sign, it's positive.

$$+2 = 2, \qquad +x = x$$

Rule 2. When adding numbers of the same sign, add them up and prefix them with the same sign as the individual numbers had.

$$+3 + (+5) = +8, \qquad 2 + 7 = 9$$

$$-4 + (-9) + (-3) = -16$$

Rule 3. When summing up a group of numbers with mixed signs, think of the process as having three steps:

1. Add up the positive numbers; add up the negative numbers as if they were positive.
2. Subtract the smaller sum from the larger sum.
3. Prefix your answer with the sign of the larger sum.

$5 - 3 + 2 - 1 \longrightarrow (5 + 2)$ and $(3 + 1) = (7)$ and (4)

7 (larger) $- 4$ (smaller) $= 3$

The answer is +3 because the positive sum (7) was larger.

$-2 + 6 - 14 + 3 - 4 \longrightarrow (6 + 3)$ and $(2 + 14 + 4) = (9)$ and (20)

20 (larger) $- 9$ (smaller) $= 11$

The answer is –11 because the negative sum (20) was larger.

Rule 4. Subtracting a positive number is the same as adding a negative number; adding a negative number is the same as subtracting a positive number. Subtracting a negative is the same as adding a positive. In other words, two negative signs make a positive sign; a positive and a negative make a negative (you visuals, work it out on the number line).

$$5 - (+3) = 5 + (-3) = 5 - 3 = 2$$

$$7 + (-12) = 7 - (+12) = 7 - 12 = -5$$

$$5 - (-4) = 5 + 4 = 9$$

Rule 5. When multiplying or dividing two numbers with the same sign, the answer is always *positive.*

$$3 \times 7 = 21, \qquad -8 \cdot -3 = 24, \qquad (12)\ (8) = 96$$

Notice the three different ways of indicating multiplication: an × sign, a "high dot" (·) between the numbers, or no sign at all. Parentheses around an expression just means to treat what's inside as a single number; we'll talk about that more a little later.

$$15/5 = 3, \qquad -9/-1 = 9, \qquad -6/-4 = 1.5$$

Rule 6. When multiplying or dividing two numbers with different signs, the answer is always *negative.*

$$3 \times -7 = -21, \qquad -8 \cdot 3 = -24, \qquad (12)(-8) = -96$$

$$-15/5 = -3, \qquad 9/-1 = -9, \qquad -6/4 = -1.5$$

Rules 5 and 6 aren't as "sensible" as some of the other rules, and the number line won't help you much with them. Just memorize.

Rule 7. With more than two numbers to be multiplied or divided, just take them pairwise, in order, and follow rules 5 and 6.

$$3 \times -2 \times -4 \times 3 = -6 \times -4 \times 3 = 24 \times 3 = 72$$

$$40/2/2/-2 = 20/2/-2 = 10/-2 = -5$$

$$6/3 \times -1/5 = 2 \times -1/5 = -2/5 = -.4$$

PROBLEMS*

1. $-5 + 4 = ?$
2. $6 - (-2) = ?$
3. $-6 - (4 + 2) = ?$
4. $3\ (-4) = ?$
5. $3 \cdot 4 = ?$
6. $(-3)(-4) = ?$
7. $(-3)(+4) = ?$
8. $(-4)(1)(2)(-3) = ?$
9. $(-a)(-b)(c)(-d) = ?$
10. $-4/-3 = ?$
11. $(-10)(3)/2 = ?$
12. $(-1)(-1)(-1)(-1)/(-1) = ?$

*The answers to all problems are set at the end of this appendix.

FRACTIONS

Rule 1. A fraction is another way of symbolizing division. A fraction means "divide the top expression (the numerator) by the bottom expression (the denominator)."

$$\frac{1}{2}=.5, \qquad \frac{4}{2}=2, \qquad \frac{4}{-2}=-2$$

$$\frac{-9}{3}=-3, \qquad \frac{6}{4-1}=2, \qquad \frac{(13-3)(7+3)}{-3+2}=\frac{100}{-1}=-100$$

Rule 2. You can turn any expression into a fraction by making the original expression the numerator, and putting a 1 into the denominator.

$$3=\frac{3}{1}, \qquad -6.2=\frac{-6.2}{1}, \qquad 3x+4=\frac{3x+4}{1}$$

Rule 3. To multiply fractions, multiply the numerators together and multiply the denominators together.

$$\frac{2}{3}\cdot\frac{1}{2}=\frac{2}{6}=.33, \qquad \frac{1}{5}\cdot 10=\frac{1}{5}\cdot\frac{10}{1}=\frac{10}{5}=2, \qquad 3xy\cdot\frac{3}{4}=\frac{3xy}{1}\cdot\frac{3}{4}=\frac{3(3xy)}{4}=\frac{9xy}{4}$$

Rule 4. Multiplying both the numerator and the denominator of a fraction by the same number doesn't change its value.

$$\frac{1}{2}=\frac{2}{4}=\frac{6}{12}=\frac{20}{40}$$

Rule 5. To divide by a fraction, invert and multiply. That is, take the fraction you're dividing by (the divisor), switch the denominator and numerator, and then multiply it times the thing you're dividing into (the dividend).

$$21\div\frac{3}{5}=21\cdot\frac{5}{3}=\frac{21}{1}\cdot\frac{5}{3}=\frac{105}{3}=35$$

$$\frac{3x}{4}\div\frac{-1}{3}=\frac{3x}{4}\cdot\frac{3}{-1}=\frac{9x}{-4}=\frac{9}{-4}\cdot\frac{x}{1}=-\frac{9}{4}x$$

Rule 6. To add or subtract fractions, they must have a common denominator; that is, their denominators must be the same. For example, you can't add 2/3 and 1/2 as they are. You have to change them to equivalent fractions with a common denominator: 2/3 = 4/6 and 1/2 = 3/6. Then add or subtract the numerators.

$$\frac{2}{3}+\frac{1}{2}=\frac{4}{6}+\frac{3}{6}=\frac{7}{6}=1\frac{1}{6}$$

$$\frac{1}{5}+\frac{1}{10}=\frac{2}{10}+\frac{1}{10}=\frac{3}{10}$$

$$\frac{5}{8}-\frac{1}{2}=\frac{5}{8}-\frac{4}{8}=\frac{1}{8}$$

$$\frac{1}{3}+\frac{1}{2}+\frac{3}{4}-\frac{1}{12}=\frac{4}{12}-\frac{6}{12}+\frac{9}{12}-\frac{1}{12}=\frac{6}{12}=\frac{1}{2}$$

PROBLEMS

1. $\frac{6}{3}=?$ **2.** $\frac{-4}{-3}=?$ **3.** $\frac{3}{4}\cdot(-1)=?$

4. $\frac{1}{2}\cdot\frac{2}{3}=?$ **5.** $5\cdot\frac{1}{2}\cdot 2\cdot\frac{6}{3}=?$ **6.** $\frac{7}{3}+\frac{1}{2}=?$

7. $\frac{1}{2}+\frac{5}{6}=?$ **8.** $\frac{7}{3}-\frac{1}{2}=?$ **9.** $5-\frac{8}{2}=?$

10. $7+\frac{1}{2}-\frac{1}{3}+\frac{1}{4}-\frac{1}{5}=?$

DECIMALS AND PERCENTS

Rule 1. Decimals indicate that the part of the number following the decimal point is a fraction with 10, 100, 1000, and so on, as the denominator. Rather than trying to find words to express the rule, let me just show you:

$$3.2=3\tfrac{2}{10},\qquad 3.25=3\tfrac{25}{100},\qquad 3.257=3\tfrac{257}{1000},\quad \text{and so on.}$$

See how it works? Not particularly complicated.

Rule 2. Some fractions, divided out, produce decimals that go on and on and on. To get rid of unneeded decimal places, we can *round off* a number. Say you have the number 1.41421 and you want to express it with just two decimal places. Should your answer be 1.41 or 1.42? The first step in deciding is to create a new number from the digits left over after you take away the ones you want to keep, with a decimal point in front of it. In our example, we keep 1.41, and the newly created number is .421. Next step:

a. If the new decimal fraction is less than .5, just throw it away; you're done with the rounding off process.
b. If the new decimal is larger than .5, throw it away but increase the last digit in your "kept" number by 1. For example, 1.41684 would round to 1.42.

c. If the new decimal is exactly .5, throw it away and look at the last digit of the "kept" number. If that last digit is even, you're all done. If it's odd, increase it by 1. For example, 5.435 rounds to 5.44; but 5.465 rounds to 5.46.

Rule 3. Percents are simply fractions of 100 (two-place decimals).

$$45\% = 45/100 = .45, \qquad 13\% = 13/100 = .13, \qquad 98\% = 98/100 = .98$$

PROBLEMS

1. Round off the following to two decimal places:
a. 3.5741 **b.** 10.1111111 **c.** 90.0054 **d.** 9.0009 **e.** 43.52500
2. Convert the following to percents:
a. .75 **b.** .7532 **c.** 1.5 **d.** 2/3 **e.** 1 − .77
3. 20% of 100 = ? **4.** .2 of 100 = ? **5.** .20 (100) = ?

EXPONENTS AND ROOTS

Rule 1. An exponent is a small number placed slightly higher than and to the right of a number or expression. For example, 3^3 has an exponent of 3; $(x - y)^2$ has an exponent of 2. An exponent tells how many times a number or expression is to be multiplied by itself.

$$5^2 = 5 \cdot 5 = 25, \qquad 10^3 = 10 \cdot 10 \cdot 10 = 1000, \qquad Y^4 = Y \cdot Y \cdot Y \cdot Y$$

Rule 2. Roots are like the opposite of exponents. You can have square roots (the opposite of an exponent of 2), cube roots (opposite of an exponent of 3), and so on. In statistics we often use square roots, and seldom any other kind, so I'm just going to talk about square roots here.

Rule 3. The square root of a number is the value that, when multiplied by itself, equals that number. For example, the square root of nine is three ($3 \cdot 3 = 9$). The instruction to compute a square root (mathematicians call it "extracting" a square root, but I think that has unfortunate associations to wisdom teeth) is a "radical" sign: $\sqrt{\ }$. You take the square root of everything that's shown under the "roof" of the radical.

$$\sqrt{9} = 3, \qquad \sqrt{8100} = 90, \qquad \sqrt{36+13} = \sqrt{49} = 7, \qquad \sqrt{36} + 13 = 6 + 13 = 19$$

When extracting a square root you have three alternatives:

a. Buy a calculator with a square root button.

b. Learn how to use a table of squares and square roots.
c. Find a sixth grader who has just studied square roots in school.

I simply cannot recommend alternative "a" too strongly, given the inconvenience of using tables and the unreliability of sixth graders.

PROBLEMS

1. $5^2 = ?$ **2.** $32^2 = ?$ **3.** $3^3 = ?$ **4.** $\sqrt{4^2} = ?$
5. $\sqrt{22{,}547} = ?$ **6.** $\sqrt{14{,}727} = ?$ **7.** $\sqrt{71.234} = ?$ **8.** $\sqrt{.0039} = ?$

ORDER OF COMPUTATION

Rule 1. When an expression is enclosed in parentheses (like this), treat what's inside like a single number. Do any operations on that expression first, before going on to what's outside the parentheses. With nested parentheses, work from the inside out.

$$4(7-2) = 4(5) = 20$$

$$9/(-4+1) = 9/(-3) = -3$$

$$12 \times (5-(6+2)) = 12 \times (5-8) = 12 \times (-3) = -36$$

With complicated fractions, treat the numerator and the denominator as if each were enclosed in parentheses. In other words, calculate the whole numerator and the whole denominator first, and then divide the numerator by the denominator:

$$\frac{3+2}{5 \cdot (7-3)} = \frac{5}{5 \cdot 4} = \frac{5}{20} = \frac{1}{4}$$

Rule 2. If you don't have parentheses to guide you, do all multiplication and division before you add or subtract.

$$5 + 3 \times 2 - 4 = 5 + (3 \times 2) - 4 = 5 + 6 - 4 = 7$$

$$\begin{aligned} 8/2 + 9 - 1 - (-2)(-5) - 5 &= (8/2) - 1 - (-2 \cdot -5) - 5 \\ &= 4 - 1 - (10) - 5 \\ &= -20 \end{aligned}$$

My algebra teacher in North Dakota taught us a device to help us remember the correct order of operation: *My Dear Aunt Sally* → *M*ultiply, *D*ivide, *A*dd, *S*ubtract.

Rule 3. Exponents and square roots are treated as if they were a single number. That means you square numbers or take square roots first of all—before adding, subtracting, multiplying, or dividing. Maybe we should treat My Dear Aunt Sally like a mean landlord, and make the rule be "Roughly Evict My Dear Aunt Sally," in order to get the Roots and Exponents first in line!

Here are some examples of how the order of computation rules work together:

$$5 - (3 \times 4) \times (8 - 2^2)(-3 + 1)/3$$

$$= 5 - (3 \times 4) \times (8 - 4)(-3 + 1)/3 \quad (\textit{exponent})$$

$$= 5 - 12 \times 6 \times -2/3 \quad (\textit{things inside parentheses})$$

$$= 5 - 72 \times -2/3 \quad (\textit{first multiplication})$$

$$= 5 - (-144)/3 \quad (\textit{next multiplication})$$

$$= 5 - (-46) \quad (\textit{division})$$

$$= 51 \quad (\textit{and the addition comes last})$$

Did you remember that subtracting a negative number is the same as adding a positive number?

$$2x - 3^2/(3 + 2) - \sqrt{25} \cdot 10 + (8 - (3 + 4))$$

$$= 2x - 9/(3 + 2) - 5 \cdot 10 + (8 - (3 + 4))$$

$$= 2x - 9/(5) - 5 \cdot 10 + (8 - 7)$$

$$= 2x - 9/(5) - 5 \cdot 10 + (1)$$

$$= 2x - 1.8 - 50 + 1$$

$$= 2x - 49.2$$

PROBLEMS

1. $3 + 2 \cdot 4 \div 5 = ?$ **2.** $3 + 2 \cdot (4 \div 5) = ?$ **3.** $(3 \div 2) \cdot 4 \div 5 = ?$

4. $(3+2 \cdot 4) \div 5 = ?$ **5.** $\dfrac{\frac{1}{2}+\frac{7}{4}}{2+\frac{1}{2}} = ?$ **6.** $\dfrac{1 \div 2 + 7 \div 4}{2 + 1 \div 2} = ?$

SUMMATION

A summation sign looks like a goat's footprint: Σ. Its meaning is pretty simple—add up what comes next. Most of the time, "what comes next" is obvious from the context. If you have a variable designated as x, with individual values x_1, x_2, x_3, and so on, then $\sum x$ refers to the sum of all those individual values.

Actually, $\sum x$ is a shorthand version of $\sum_{i=1}^{N} x$, which means that there are N individual x's. Each x is called x_i, and the values of i run from 1 to N. When $i = 10$ and $N = 50$, x_i would be the tenth in a set of 50 variables; $\sum_{i=1}^{N} x$, would mean to find the sum of all 50 of them. For our purposes, a simple $\sum x$ says the same thing and we'll just use that.

There are a few rules that you should know, though, about doing the summation. Let's look at an example. Five people take a pop quiz, and their scores are 10, 10, 8, 12, and 10. In other words, $x_1 = 10$, $x_2 = 10$, $x_3 = 8$, $x_4 = 12$, and $x_5 = 10$. $\sum x = 50$. What about $\sum x^2$? Well, that would be $100 + 100 + 64 + 144 + 100$. $\sum x^2 = 508$.

Now, having gotten that far, does it make sense to you that

$$\sum x^2 \neq (\sum x)^2$$

This is a key idea, and it has to do with the order of computation. $(\sum x)^2$ is read "sum of x, quantity squared," and the parentheses mean that you add up all the x's first, and square the sum: $(\sum x)^2 = 50^2 = 2500$.

Now, just to make things interesting, we'll throw in another variable. Let y stand for scores on another quiz. $y_1 = 4$, $y_2 = 5$, $y_3 = 6$, $y_4 = 5$, and $y_5 = 4$. $\sum y = 24$, $\sum y^2 = 118$, and $\sum (y)^2 = 576$. And we have some new possibilities:

$$\sum x + \sum y,\quad \sum x^2 + \sum y^2,\quad \sum (x + y),\quad \sum (x^2 + y^2),\quad \sum (x + y)^2,\ (\sum (x + y))^2$$

See if you can figure out these values on your own, and then we'll go through each one.

$\sum x + \sum y$	Add up the x values, add up the y values, add them together: 70.
$\sum x^2 + \sum y^2$	Add up the squared x values, add up the squared y values, add them together: 626.
$\sum (x + y)$	Add each x, y pair together, and add up the sums: $14 + 15 + 14 + 18 + 15 = 70$. Yup, $\sum x + \sum y = \sum (x + y)$. Every time.
$\sum (x^2 + y^2)$	Square an x and add it to its squared y partner; then add up the sums: $116 + 125 + 100 + 169 + 116 = 626$. $\sum (x^2 + y^2) = \sum x^2 + \sum y^2$.
$\sum (x + y)^2$	Add each x, y pair together, square the sums, and add them up: 1166.
$(\sum (x + y))^2$	Did those double parentheses throw you? Use them like a road map, to tell you where to go first. Starting from the inside, you add each x, y pair together. Find the sum of the pairs, and last of all square that sum: 4900.

PROBLEMS

Use these values to solve the following problems:

x	y
1	5
2	4
3	3
4	2
5	1

1. $\sum x + \sum y$ **2.** $\sum x^2 + \sum y^2$ **3.** $\sum (x + y)$ **4.** $\sum (x^2 + y^2)$ **5.** $\sum (x + y)^2$
6. $(\sum (x + y))^2$

EQUATIONS

An equation is two expressions joined by an equal sign. Not surprisingly, the value of the part in front of the equal sign is exactly equal to the value of the part after the equal sign.

Rule 1. Adding or subtracting the same number from each side of an equation is acceptable; the two sides will still be equivalent.

$5 + 3 = 9 - 1$ $\quad$ $5 + 3\ [+5] = 9 - 1\ [+5]$ $\quad$ $5 + 3\ [-5] = 9 - 1\ [-5]$

$13 = 13$ $\quad$ $3 = 3$ $\quad$ $3 = 3$

$6/4 + 1/2 = 2$ $\quad$ $6/4 + 1/2\ [+5] = 2\ [+5]$ $\quad$ $6/4 + 1/2\ [-5] = 2\ [-5]$

$2 = 2$ $\quad$ $7 = 7$ $\quad$ $-3 = -3$

$12 - 2 = (2)(5)$ $\quad$ $12 - 2\ [+5] = (2)(5) + [+5]$ $\quad$ $12 - 2\ [-5] = (2)(5) - [+5]$

$10 = 10$ $\quad$ $15 = 15$ $\quad$ $5 = 5$

Notice that I've used brackets in these examples just to indicate the numbers that we are adding or subtracting. The brackets have no mathematical significance; they are just there to clarify what we are doing.

Rule 2. If you add or subtract a number from one side of an equation, you must add or subtract it from the other side as well if the equation is to *balance,* that is, if both sides are to remain equal.

$$8 - 2 = 3 + 3, \qquad 8 - 2 + 2 = 3 + 3 + 2, \qquad 8 = 8$$

$$2x + 7 = 35, \qquad 2x + 7 - 7 = 35 - 7, \qquad 2x = 28$$

Rule 3. If you multiply or divide one side of an equation by some number, you must multiply or divide the other side by the same number. You can't multiply or divide just one part of each side; you have to multiply or divide the whole thing.

$$3 + 2 - 1 = 7 - 5 + 2$$

Multiply both sides by 6:

$$6(3 + 2 - 1) = 6(7 - 5 + 2)$$
$$6(4) = 6(4)$$
$$24 = 24$$

Look what would happen if you multiplied just one of the numbers on each side by 6:

$$6(3) + 2 - 1 = 6(7) - 5 + 2$$
$$18 + 2 - 1 = 42 - 5 + 2$$
$$19 = 39$$

PROBLEMS

1. Using the equation $2 \cdot 3 + 4 = 10$,
   - **a.** demonstrate that you can add the same number to each side of the equation without changing its balance.
   - **b.** show the same thing, using subtraction.
   - **c.** multiply both sides of the equation by 2.
   - **d.** divide both sides of the equation by 2.
2. Use addition and/or subtraction to solve these equations:
   - **a.** $5 + x = 3 - 7$
   - **b.** $x - 3 = 10$
   - **c.** $x - 3 + 2 = 8/4$
3. Use multiplication and/or division to solve these equations:
   - **a.** $3x = 12$
   - **b.** $x/4 = 3$
   - **c.** $2x - 7 = 8$

Writing out these kinds of rules is a lot like eating hot buttered popcorn: it's hard to know when to quit. And, as with popcorn, it's a lot better to quit too soon than to quit too late; the former leaves you ready for more tomorrow, while the latter can make you swear off the stuff for months. We could go on and on here, and end up with the outline for a freshman math course, but that's not our purpose. These rules will allow you to do all the math in this book and a great deal of the math in more advanced statistics courses. So let's get going on the fun part!

ANSWERS

Positive and Negative Numbers

1. –1 **2.** 8 **3.** –12
4. –12 **5.** 12 **6.** 12
7. –12 **8.** 24 **9.** –(*abcd*)
10. 4/3 or $1\frac{1}{3}$ **11.** –15 **12.** –1

Fractions

1. 2 **2.** $1\frac{1}{3}$

3. $-\left(\frac{3}{4}\right)$ **4.** $\frac{2}{6}=\frac{1}{3}$

5. $\frac{5}{1}\cdot\frac{1}{2}\cdot\frac{2}{1}\cdot\frac{6}{3}=\frac{60}{6}=10$ **6.** $\frac{7}{3}\cdot\frac{2}{1}=\frac{14}{3}=4\frac{2}{3}$

7. $\frac{3}{6}\cdot\frac{5}{6}=\frac{8}{6}=1\frac{2}{6}=1\frac{1}{3}$ **8.** $\frac{14}{6}-\frac{3}{6}=\frac{11}{6}=1\frac{5}{6}$

9. $\frac{10}{2}-\frac{8}{2}=\frac{2}{2}=1$ **10.** $7+\frac{30}{60}-\frac{20}{60}+\frac{15}{60}-\frac{12}{60}=7\frac{13}{60}$

Decimals and Percents

1a. 3.57 **1b.** 10.11 **1c.** 90.01 **1d.** 9.00 **1e.** 43.52
2a. 75% **2b.** 75% **2c.** 150% **2d.** 67% **2e.** 33%
3. 20 **4.** 20 **5.** 20

Exponents and Roots

1. 25 **2.** 1024 **3.** 9 **4.** 4
5. 150.16 **6.** 121.35 **7.** 8.44 **8.** .06

Order of Computation

1. 4.6 **2.** 4.6 **3.** 4 **4.** 4

5. $\frac{\frac{9}{4}}{\frac{5}{2}}=\frac{9}{4}\cdot\frac{2}{5}=\frac{18}{20}=\frac{9}{10}$ **6.** *Exactly the same as answer* 5

Summation

1. 30 **2.** 110 **3.** 30 **4.** 110 **5.** 216 **6.** 900

Equations

1a. $2 \cdot 3 + 4 = 10$ $2 \cdot 3 + 4 + 100 = 10 + 100$ $10 = 10$ $110 = 110$

1b. $2 \cdot 3 + 4 = 10$ $2 \cdot 3 + 4 - 100 = 10 - 100$ $10 = 10$ $-90 = -90$

1c. $2 \cdot 3 + 4 = 10$ $2(2 \cdot 3 + 4) = 2(10)$ $10 = 10$ $2(10) = 2(10)$

1d. $2 \cdot 3 + 4 = 10$ $\frac{2 \cdot 3 + 4}{2} = \frac{10}{2}$

$10 = 10$ $\frac{10}{2} = \frac{10}{2}$

2a. $x - 3 = 10$

$x - 3 + 3 = 10 + 3$

$x = 13$

2b. $5 + x = 3 - 7$

$5 + x - 5 = 3 - 7 - 5$

$x = -9$

2c. $x - 3 + 2 = 8/4$

$x - 5 = 2$

$x - 5 + 5 = 2$

$x = 2$

3a. $3x = 12$

$\frac{3x}{3} = \frac{12}{3}$

$x = 4$

3b. $\frac{x}{4} = 3$

$\left(\frac{x}{4}\right) \cdot 4 = 3 \cdot 4$

$x = 12$

3c. $2x - 7 = 8$

$2x - 7 + 7 = 8 + 7$

$2x = 15$

$x = 7\frac{1}{2}$

APPENDIX B

Proportions of Area under the Standard Normal Curve

z	0 z	0 z	z	0 z	0 z	z	0 z	0 z
0.00	.0000	.5000	0.17	.0675	.4325	0.34	.1331	.3669
0.01	.0040	.4960	0.18	.0714	.4286	0.35	.1368	.3632
0.02	.0080	.4920	0.19	.0753	.4247	0.36	.1406	.3594
0.03	.0120	.4880	0.20	.0793	.4207	0.37	.1443	.3557
0.04	.0160	.4840	0.21	.0832	.4168	0.38	.1480	.3520
0.05	.0199	.4801	0.22	.0871	.4129	0.39	.1517	.3483
0.06	.0239	.4761	0.23	.0910	.4090	0.40	.1554	.3446
0.07	.0279	.4721	0.24	.0948	.4052	0.41	.1591	.3409
0.08	.0319	.4681	0.25	.0987	.4013	0.42	.1628	.3372
0.09	.0359	.4641	0.26	.1026	.3974	0.43	.1664	.3336
0.10	.0398	.4602	0.27	.1064	.3936	0.44	.1700	.3300
0.11	.0438	.4562	0.28	.1103	.3897	0.45	.1736	.3264
0.12	.0478	.4522	0.29	.1141	.3859	0.46	.1772	.3228
0.13	.0517	.4483	0.30	.1179	.3821	0.47	.1808	.3192
0.14	.0557	.4443	0.31	.1217	.3783	0.48	.1844	.3156
0.15	.0596	.4404	0.32	.1255	.3745	0.49	.1879	.3121
0.16	.0636	.4364	0.33	.1293	.3707	0.50	.1915	.3085

Source: Runyon and Haber, ***Fundamentals of Behavioral Statistics,*** 2 ed., 1971, Addison-Wesley, Reading, Mass.

APPENDIX B (continued)

z	0 z	0 z	z	0 z	0 z	z	0 z	0 z
0.51	.1950	.3050	0.89	.3133	.1867	1.27	.3980	.1020
0.52	.1985	.3015	0.90	.3159	.1841	1.28	.3997	.1003
0.53	.2019	.2981	0.91	.3186	.1814	1.29	.4015	.0985
0.54	.2054	.2946	0.92	.3212	.1788	1.30	.4032	.0968
0.55	.2088	.2912	0.93	.3238	.1762	1.31	.4049	.0951
0.56	.2123	.2877	0.94	.3264	.1736	1.32	.4066	.0934
0.57	.2157	.2843	0.95	.3289	.1711	1.33	.4082	.0918
0.58	.2190	.2810	0.96	.3315	.1685	1.34	.4099	.0901
0.59	.2224	.2776	0.97	.3340	.1660	1.35	.4115	.0885
0.60	.2257	.2743	0.98	.3365	.1635	1.36	.4131	.0869
0.61	.2291	.2709	0.99	.3389	.1611	1.37	.4147	.0853
0.62	.2324	.2676	1.00	.3413	.1587	1.38	.4162	.0838
0.63	.2357	.2643	1.01	.3438	.1562	1.39	.4177	.0823
0.64	.2389	.2611	1.02	.3461	.1539	1.40	.4192	.0808
0.65	.2422	.2578	1.03	.3485	.1515	1.41	.4207	.0793
0.66	.2454	.2546	1.04	.3508	.1492	1.42	.4222	.0778
0.67	.2486	.2514	1.05	.3531	.1469	1.43	.4236	.0764
0.68	.2517	.2483	1.06	.3554	.1446	1.44	.4251	.0749
0.69	.2549	.2451	1.07	.3577	.1423	1.45	.4265	.0735
0.70	.2580	.2420	1.08	.3599	.1401	1.46	.4279	.0721
0.71	.2611	.2389	1.09	.3621	.1379	1.47	.4292	.0708
0.72	.2642	.2358	1.10	.3643	.1357	1.48	.4306	.0694
0.73	.2673	.2327	1.11	.3665	.1335	1.49	.4319	.0681
0.74	.2704	.2296	1.12	.3686	.1314	1.50	.4332	.0668
0.75	.2734	.2266	1.13	.3708	.1292	1.51	.4345	.0655
0.76	.2764	.2236	1.14	.3729	.1271	1.52	.4357	.0643
0.77	.2794	.2206	1.15	.3749	.1251	1.53	.4370	.0630
0.78	.2823	.2177	1.16	.3770	.1230	1.54	.4382	.0618
0.79	.2852	.2148	1.17	.3790	.1210	1.55	.4394	.0606
0.80	.2881	.2119	1.18	.3810	.1190	1.56	.4406	.0594
0.81	.2910	.2090	1.19	.3830	.1170	1.57	.4418	.0582
0.82	.2939	.2061	1.20	.3849	.1151	1.58	.4429	.0571
0.83	.2967	.2033	1.21	.3869	.1131	1.59	.4441	.0559
0.84	.2995	.2005	1.22	.3888	.1112	1.60	.4452	.0548
0.85	.3023	.1977	1.23	.3907	.1093	1.61	.4463	.0537
0.86	.3051	.1949	1.24	.3925	.1075	1.62	.4474	.0526
0.87	.3078	.1922	1.25	.3944	.1056	1.63	.4484	.0516
0.88	.3106	.1894	1.26	.3962	.1038	1.64	.4495	.0505

APPENDIX B (continued)

z	0 z	0 z	z	0 z	0 z	z	0 z	0 z
1.65	.4505	.0495	2.03	.4788	.0212	2.41	.4920	.0080
1.66	.4515	.0485	2.04	.4793	.0207	2.42	.4922	.0078
1.67	.4525	.0475	2.05	.4798	.0202	2.43	.4925	.0075
1.68	.4535	.0465	2.06	.4803	.0197	2.44	.4927	.0073
1.69	.4545	.0455	2.07	.4808	.0192	2.45	.4929	.0071
1.70	.4554	.0446	2.08	.4812	.0188	2.46	.4931	.0069
1.71	.4564	.0436	2.09	.4817	.0183	2.47	.4932	.0068
1.72	.4573	.0427	2.10	.4821	.0179	2.48	.4934	.0066
1.73	.4582	.0418	2.11	.4826	.0174	2.49	.4936	.0064
1.74	.4591	.0409	2.12	.4830	.0170	2.50	.4938	.0062
1.75	.4599	.0401	2.13	.4834	.0166	2.51	.4940	.0060
1.76	.4608	.0392	2.14	.4838	.0162	2.52	.4941	.0059
1.77	.4616	.0384	2.15	.4842	.0158	2.53	.4943	.0057
1.78	.4625	.0375	2.16	.4846	.0154	2.54	.4945	.0055
1.79	.4633	.0367	2.17	.4850	.0150	2.55	.4946	.0054
1.80	.4641	.0359	2.18	.4854	.0146	2.56	.4948	.0052
1.81	.4649	.0351	2.19	.4857	.0143	2.57	.4949	.0051
1.82	.4656	.0344	2.20	.4861	.0139	2.58	.4951	.0049
1.83	.4664	.0336	2.21	.4864	.0136	2.59	.4952	.0048
1.84	.4671	.0329	2.22	.4868	.0132	2.60	.4953	.0047
1.85	.4678	.0322	2.23	.4871	.0129	2.61	.4955	.0045
1.86	.4686	.0314	2.24	.4875	.0125	2.62	.4956	.0044
1.87	.4693	.0307	2.25	.4878	.0122	2.63	.4957	.0043
1.88	.4699	.0301	2.26	.4881	.0119	2.64	.4959	.0041
1.89	.4706	.0294	2.27	.4884	.0116	2.65	.4960	.0040
1.90	.4713	.0287	2.28	.4887	.0113	2.66	.4961	.0039
1.91	.4719	.0281	2.29	.4890	.0110	2.67	.4962	.0038
1.92	.4726	.0274	2.30	.4893	.0107	2.68	.4963	.0037
1.93	.4732	.0268	2.31	.4896	.0104	2.69	.4964	.0036
1.94	.4738	.0262	2.32	.4898	.0102	2.70	.4965	.0035
1.95	.4744	.0256	2.33	.4901	.0099	2.71	.4966	.0034
1.96	.4750	.0250	2.34	.4904	.0096	2.72	.4967	.0033
1.97	.4756	.0244	2.35	.4906	.0094	2.73	.4968	.0032
1.98	.4761	.0239	2.36	.4909	.0091	2.74	.4969	.0031
1.99	.4767	.0233	2.37	.4911	.0089	2.75	.4970	.0030
2.00	.4772	.0228	2.38	.4913	.0087	2.76	.4971	.0029
2.01	.4778	.0222	2.39	.4916	.0084	2.77	.4972	.0028
2.02	.4783	.0217	2.40	.4918	.0082	2.78	.4973	.0027

APPENDIX B (continued)

z	0 z (area between 0 and z)	0 z (area beyond z)	z	0 z (area between 0 and z)	0 z (area beyond z)	z	0 z (area between 0 and z)	0 z (area beyond z)
2.79	.4974	.0026	2.98	.4986	.0014	3.17	.4992	.0008
2.80	.4974	.0026	2.99	.4986	.0014	3.18	.4993	.0007
2.81	.4975	.0025	3.00	.4987	.0013	3.19	.4993	.0007
2.82	.4976	.0024	3.01	.4987	.0013	3.20	.4993	.0007
2.83	.4977	.0023	3.02	.4987	.0013	3.21	.4993	.0007
2.84	.4977	.0023	3.03	.4988	.0012	3.22	.4994	.0006
2.85	.4978	.0022	3.04	.4988	.0012	3.23	.4994	.0006
2.86	.4979	.0021	3.05	.4989	.0011	3.24	.4994	.0006
2.87	.4979	.0021	3.06	.4989	.0011	3.25	.4994	.0006
2.88	.4980	.0020	3.07	.4989	.0011	3.30	.4995	.0005
2.89	.4981	.0019	3.08	.4990	.0010	3.35	.4996	.0004
2.90	.4981	.0019	3.09	.4990	.0010	3.40	.4997	.0003
2.91	.4982	.0018	3.10	.4990	.0010	3.45	.4997	.0003
2.92	.4982	.0018	3.11	.4991	.0009	3.50	.4998	.0002
2.93	.4983	.0017	3.12	.4991	.0009	3.60	.4998	.0002
2.94	.4984	.0016	3.13	.4991	.0009	3.70	.4999	.0001
2.95	.4984	.0016	3.14	.4992	.0008	3.80	.4999	.0001
2.96	.4985	.0015	3.15	.4992	.0008	3.90	.49995	.00005
2.97	.4985	.0015	3.16	.4992	.0008	4.00	.49997	.00003

APPENDIX C

Pearson Product–Moment Correlation Coefficient Values

df = $N-2$	Level of Significance for a Nondirectional (Two-tailed) Test				
	.10	.05	.02	.01	.001
1	.9877	.9969	.9995	.9999	1.0000
2	.9000	.9500	.9800	.9900	.9990
3	.8054	.8783	.9343	.9587	.9912
4	.7293	.8114	.8822	.9172	.9741
5	.6694	.7545	.8329	.8745	.9507
6	.6215	.7067	.7887	.8343	.9249
7	.5822	.6664	.7498	.7977	.8982
8	.5494	.6319	.7155	.7646	.8721
9	.5214	.6021	.6851	.7348	.8471
10	.4973	.5760	.6581	.7079	.8233
11	.4762	.5529	.6339	.6835	.8010
12	.4575	.5324	.6120	.6614	.7800
13	.4409	.5139	.5923	.6411	.7603
14	.4259	.4973	.5742	.6226	.7420
15	.4124	.4821	.5577	.6055	.7246

Source: This table is taken from Table VII of Fisher and Yates, *Statistical Tables for Biological, Agricultural, and Medical Research,* published by Longman Group Ltd., London (previously published by Oliver and Boyd, Ltd., Edinburgh), and by permission of the authors and publishers.

APPENDIX C (continued)

	Level of Significance for a Nondirectional (Two-tailed) Test				
df = $N - 2$	.10	.05	.02	.01	.001
16	.4000	.4683	.5425	.5897	.7084
17	.3887	.4555	.5285	.5751	.6932
18	.3783	.4438	.5155	.5614	.6787
19	.3687	.4329	.5034	.5487	.6652
20	.3598	.4227	.4921	.5368	.6524
25	.3233	.3809	.4451	.4869	.5974
30	.2960	.3494	.4093	.4487	.5541
35	.2746	.3246	.3810	.4182	.5189
40	.2573	.3044	.3578	.3932	.4896
45	.2428	.2875	.3384	.3721	.4648
50	.2306	.2732	.3218	.3541	.4433
60	.2108	.2500	.2948	.3248	.4078
70	.1954	.2319	.2737	.3017	.3799
80	.1829	.2172	.2565	.2830	.3568
90	.1726	.2050	.2422	.2673	.3375
100	.1638	.1946	.2301	.2540	.3211

APPENDIX D

Table of Random Numbers

Using the Random Number Table to Draw a Sample

Step 1. Define your population, for example, all the fifth grade students in the six elementary schools in Kokomo, Indiana.

Step 2. List all the members of the population. In our example, you would have to go to the individual schools or to the Board of Education and get this information. (This is the hardest step.)

Step 3. Assign a number to each member of the population. 1, 2, 3, 4, 5, and on out through the last student on your list.

Step 4. Decide on the size of your sample. This will depend on all sorts of things: the kind of experiment you plan to do, the consequences of drawing a wrong conclusion (the likelihood of error goes down as the sample size goes up), the amount of money available. Let's say you decide to draw a sample of 50.

Step 5. The numbers in the random number table are grouped into 5-digit sets. These groupings are merely for your convenience, to help you keep your place. Select a column of as many adjacent numbers as the number of digits in your sample size. Since your sample has 50, you'll need to use 2-digit columns. For a sample of 200, you would use sets of 3 adjacent columns. Your columns of digits can come from any grouping, from anywhere in that grouping, or can even span across groupings if you wish. Decide where you want to enter your set of columns by closing your eyes and putting your finger on the column.

Step 6. The first number you find, if it is 50 or less, is the number of the first child in your sample. Find the child with that number, and record his or her name. The next number identifies the next child. Keep on identifying sample members this way until you have 50 of them, discarding any numbers that are repeated or that are above 50.

Randomly Assigning Subjects to Treatment Groups

Suppose that you have 30 subjects and that you want to assign 10 subjects to each of three treatment groups:

1. In the table of random numbers, make a blind selection of a 2-digit column (because your total *N*, 30, has 2 digits).
2. List the first 30 numbers from that column on a piece of paper (or on your word processor).
3. Write or type the names of your 30 subjects in a second list, pairing each name with one of the numbers from step 2.
4. Rearrange the names on the list so that they are in numerical order according to the numbers assigned to them.
5. Put the first 10 names into treatment group 1, the next 10 into treatment group 2, and the last 10 into treatment group 3.

00	54463	22662	65905	70639	79365	67382	29085	69831	47058	08186
01	15389	85205	18850	39226	42249	90669	96325	23248	60933	26927
02	85941	40756	82414	02015	13858	78030	16269	65978	01385	15345
03	61149	69440	11286	88218	58925	03638	52862	62733	33451	77455
04	05219	81619	10651	67079	92511	59888	84502	72095	83463	75577
05	41417	98326	87719	92294	46614	50948	64886	20002	97365	30976
06	28357	94070	20652	35774	16249	75019	21145	05217	47286	76305
07	17783	00015	10806	83091	91530	36466	39981	62481	49177	75779
08	40950	84820	29881	85966	62800	70326	84740	62660	77379	90279
09	82995	64157	66164	41180	10089	41757	78258	96488	88629	37231
10	96754	17676	55659	44105	47361	34833	86679	23930	53249	27083
11	34357	88040	53364	71726	45690	66334	60332	22554	90600	71113
12	06318	37403	49927	57715	50423	67372	63116	48888	21505	80182
13	62111	52820	07243	79931	89292	84767	85693	73947	22278	11551
14	47534	09243	67879	00544	23410	12740	02540	54440	32949	13491
15	98614	75993	84460	62846	59844	14922	48730	73443	48167	34770
16	24856	03648	44898	09351	98795	18644	39765	71058	90368	44104
17	96887	12479	80621	66223	86085	78285	02432	53342	42846	94771
18	90801	21472	42815	77408	37390	76766	52615	32141	30268	18106
19	55165	77312	83666	36028	28420	70219	81369	41943	47366	41067
20	75884	12952	84318	95108	72305	64620	91318	89872	45375	85436
21	16777	37116	58550	42958	21460	43910	01175	87894	81378	10620
22	46230	43877	80207	88877	89380	32992	91380	03164	98656	59337
23	42902	66892	46134	01432	94710	23474	20423	60137	60609	13119
24	81007	00333	39693	28039	10154	95425	39220	19774	31782	49037
25	68089	01122	51111	72373	06902	74373	96199	97017	41273	21546
26	20411	67081	89950	16944	93054	87687	96693	87236	77054	33848
27	58212	13160	06468	15718	82627	76999	05999	58680	96739	63700
28	70577	42866	24969	61210	76046	67699	42054	12696	93758	03283
29	94522	74358	71659	62038	79643	79169	44741	05437	39038	13163
30	42626	86819	85651	88678	17401	03252	99547	32404	17918	62880
31	16051	33763	57194	16752	54450	19031	58580	47629	54132	60631
32	08244	27647	33851	44705	94211	46716	11738	55784	95374	72655
33	59497	04392	09419	89964	51211	04894	72882	17805	21896	83864
34	97155	13428	40293	09985	58434	01412	69124	82171	59058	82859
35	98409	66162	95763	47420	20792	61527	20441	39435	11859	41567
36	45476	84882	65109	96597	25930	66790	65706	61203	53634	22557
37	89300	69700	50741	30329	11658	23166	05400	66669	48708	03887
38	50051	95137	91631	66315	91428	12275	24816	68091	71710	33258
39	31753	85178	31310	89642	98364	02306	24617	09609	83942	22716
40	79152	53829	77250	20190	56535	18760	69942	77448	33278	48805
41	44560	38750	83635	56540	64900	42912	13953	79149	18710	68618
42	68328	83378	63369	71381	39564	05615	42451	64559	97501	65747
43	46939	38689	58625	08342	30459	85863	20781	09284	26333	91777
44	83544	86141	15707	96256	23068	13782	08467	89469	93842	55349
45	91621	00881	04900	54224	46177	55309	17852	27491	89415	23466
46	91896	67126	04151	03795	59077	11848	12630	98375	52068	60142
47	55751	62515	21108	80830	02263	29303	37204	96926	30506	09808
48	85156	87689	95493	88842	00664	55017	55539	17771	69448	87530
49	07521	56898	12236	60277	39102	62315	12239	07105	11844	01117

00	59391	58030	52098	82718	87024	82848	04190	96574	90464	29065
01	99567	76364	77204	04615	27062	96621	43918	01896	83991	51141
02	10363	97518	51400	25670	98342	61891	27101	37855	06235	33316
03	86859	19558	64432	16706	99612	59798	32803	67708	15297	28612
04	11258	24591	36863	55368	31721	94335	34936	02566	80972	08188
05	95068	88628	35911	14530	33020	80428	39936	31855	34334	64865
06	54463	47237	73800	91017	36239	71824	83671	39892	60518	37092
07	16874	62677	57412	13215	31389	62233	80827	73917	82802	84420
08	92494	63157	76593	91316	03505	72389	96363	52887	01087	66091
09	15669	56689	35682	40844	53256	81872	35213	09840	34471	74441
10	99116	75486	84989	23476	52967	67104	39495	39100	17217	74073
11	15696	10703	65178	90637	63110	17622	53988	71087	84148	11670
12	97720	15369	51269	69620	03388	13699	33423	67453	43269	56720
13	11666	13841	71681	98000	35979	39719	81899	07449	47985	46967
14	71628	73130	78783	75691	41632	09847	61547	18707	85489	69944
15	40501	51089	99943	91843	41995	88931	73631	69361	05375	15417
16	22518	55576	98215	82068	10798	86211	36584	67466	69373	40054
17	75112	30485	62173	02132	14878	92879	22281	16783	86352	00077
18	80327	02671	98191	84342	90813	49268	95441	15496	20168	09271
19	60251	45548	02146	05597	48228	81366	34598	72856	66762	17002
20	57430	82270	10421	05540	43648	75888	66049	21511	47676	33444
21	73528	39559	34434	88596	54086	71693	43132	14414	79949	85193
22	26001	65959	70769	64721	86413	33475	42740	06175	82758	66248
23	78388	16638	09134	59880	63806	48472	39318	35434	24057	74739
24	12477	09965	96657	57994	59439	76330	24596	77515	09577	91871
25	83266	32883	42451	15579	38155	29793	40914	65990	16255	17777
26	76970	80870	10207	30516	70153	74798	39357	09054	73579	92359
27	37074	65198	44785	68624	98336	84481	97610	78735	46703	98265
28	83712	06514	30101	78295	54656	85417	43189	60048	72781	72606
29	20287	56862	69727	94443	64936	08366	27227	05158	50326	59566
30	74261	32592	86538	27041	65172	85532	07571	80609	39285	65340
31	64081	49863	08478	96001	18888	14810	70545	89755	59064	07210
32	05617	75818	47750	67814	29575	10526	66192	44464	27058	40467
33	26793	74951	95466	74307	13330	42664	85515	20632	05497	33625
34	65988	72850	48737	54719	52056	01596	03845	35067	03134	70322
35	27366	42271	44300	73399	21105	03280	73457	43093	05192	48657
36	56760	10909	98147	34736	33863	95256	12731	66598	50771	83665
37	72880	43338	93643	58904	59543	23943	11231	83268	65938	81581
38	77888	38100	03062	58103	47961	83841	25878	23746	55903	44115
39	28440	07819	21580	51459	47971	29882	13990	29226	23608	15873
40	63525	94441	77033	12147	51054	49955	58312	76923	96071	05813
41	47606	93410	16359	89033	89696	47231	64498	31776	05383	39902
42	52669	45030	96279	14709	52372	87832	02735	50803	72744	88208
43	16738	60159	07425	62369	07515	82721	37875	71153	21315	00132
44	59348	11695	45751	15865	74739	05572	32688	20271	65128	14551
45	12900	71775	29845	60774	94924	21810	38636	33717	67598	82521
46	75086	23537	49939	33595	13484	97588	28617	17979	70749	35234
47	99495	51434	29181	09993	38190	42553	68922	52125	91077	40197
48	26075	31671	45386	36583	93459	48599	52022	41330	60651	91321
49	13636	93596	23377	51133	95126	61496	42474	45141	46660	42338

50	64249	63664	39652	40646	97306	31741	07294	84149	46797	82487
51	26538	44249	04050	48174	65570	44072	40192	51153	11397	58212
52	05845	00512	78630	55328	18116	69296	91705	86224	29503	57071
53	74897	68373	67359	51014	33510	83048	17056	72506	82949	54600
54	20872	54570	35017	88132	25730	22626	86723	91691	13191	77212
55	31432	96156	89177	75541	81355	24480	77243	76690	42507	84362
56	66890	61505	01240	00660	05873	13568	76082	79172	57913	93448
57	48194	57790	79970	33106	86904	48119	52503	24130	72824	21627
58	11303	87118	81471	52936	08555	28420	49416	44448	04269	27029
59	54374	57325	16947	45356	78371	10563	97191	53798	12693	27928
60	64852	34421	61046	90849	13966	39810	42699	21753	76192	10508
61	16309	20384	09491	91588	97720	89846	30376	76970	23063	35894
62	42587	37065	24526	72602	57589	98131	37292	05967	26002	51945
63	40177	98590	97161	41682	84533	67588	62036	49967	01990	72308
64	82309	76128	93965	26743	24141	04838	40254	26065	07938	76236
65	79788	68243	59732	04257	27084	14743	17520	95401	55811	76099
66	40538	79000	89559	25026	42274	23489	34502	75508	06059	86682
67	64016	73598	18609	73150	62463	33102	45205	87440	96767	67042
68	49767	12691	17903	93871	99721	79109	09425	26904	07419	76013
69	76974	55108	29795	08404	82684	00497	51126	79935	57450	55671
70	23854	08480	85983	96025	50117	64610	99425	62291	86943	21541
71	68973	70551	25098	78033	98573	79848	31778	29555	61446	23037
72	36444	93600	65350	14971	25325	00427	52073	64280	18847	24768
73	03003	87800	07391	11594	21196	00781	32550	57158	58887	73041
74	17540	26188	36647	78386	04558	61463	57842	90382	77019	24210
75	38916	55809	47982	41968	69760	79422	80154	91486	19180	15100
76	64288	19843	69122	42502	48508	28820	59933	72998	99942	10515
77	86809	51564	38040	39418	49915	19000	58050	16899	79952	57849
78	99800	99566	14742	05028	30033	94889	53381	23656	75787	59223
79	92345	31890	95712	08279	91794	94068	49337	88674	35355	12267
80	90363	65162	32245	82279	79256	80834	06088	99462	56705	06118
81	64437	32242	48431	04835	39070	59702	31508	60935	22390	52246
82	91714	53662	28373	34333	55791	74758	51144	18827	10704	76803
83	20902	17646	31391	31459	33315	03444	55743	74701	58851	27427
84	12217	86007	70371	52281	14510	76094	96579	54853	78339	20839
85	45177	02863	42307	53571	22532	74921	17735	42201	80540	54721
86	28325	90814	08804	52746	47913	54577	47525	77705	95330	21866
87	29019	28776	56116	54791	64604	08815	46049	71186	34650	14994
88	84979	81353	56219	67062	26146	82567	33122	14124	46240	92973
89	50371	26347	48513	63915	11158	25563	91915	18431	92978	11591
90	53422	06825	69711	67950	64716	18003	49581	45378	99878	61130
91	67453	35651	89316	41620	32048	70225	47597	33137	31443	51445
92	07294	85353	74819	23445	68237	07202	99515	62282	53809	26685
93	79544	00302	45338	16015	66613	88968	14595	63836	77716	79596
94	64144	85442	82060	46471	24162	39500	87351	36637	42833	71875
95	90919	11883	58318	00042	52402	28210	34075	33272	00840	73268
96	06670	57353	86275	92276	77591	46924	60839	55437	03183	13191
97	36634	93976	52062	83678	41256	60948	18685	48992	19462	96062
98	75101	72891	85745	67106	26010	62107	60885	37503	55461	71213
99	05112	71222	72654	51583	05228	62056	57390	42746	39272	96659

50	32847	31282	03345	89593	69214	70381	78285	20054	91018	16742
51	16916	00041	30236	55023	14253	76582	12092	86533	92426	37655
52	66176	34047	21005	27137	03191	48970	64625	22394	39622	79085
53	46299	13335	12180	16861	38043	59292	62675	63631	37020	78195
54	22847	47839	45385	23289	47526	54098	45683	55849	51575	64689
55	41851	54160	92320	69936	34803	92479	33399	71160	64777	83378
56	28444	59497	91586	95917	68553	28639	06455	34174	11130	91994
57	47520	62378	98855	83174	13088	16561	68559	26679	06238	51254
58	34978	63271	13142	82681	05271	08822	06490	44984	49307	62717
59	37404	80416	69035	92980	49486	74378	75610	74976	70056	15478
60	32400	65482	52099	53676	74648	94148	65095	69597	52771	71551
61	89262	86332	51718	70663	11623	29834	79820	73002	84886	03591
62	86866	09127	98021	03871	27789	58444	44832	36505	40672	30180
63	90814	14833	08759	74645	05046	94056	99094	65091	32663	73040
64	19192	82756	20553	58446	55376	88914	75096	26119	83898	43816
65	77585	52593	56612	95766	10019	29531	73064	20953	53523	58136
66	23757	16364	05096	03192	62386	45389	85332	18877	55710	96459
67	45989	96257	23850	26216	23309	21526	07425	50254	19455	29315
68	92970	94243	07316	41467	64837	52406	25225	51553	31220	14032
69	74346	59596	40088	98176	17896	86900	20249	77753	19099	48885
70	87646	41309	27636	45153	29988	94770	07255	70908	05340	99751
71	50099	71038	45146	06146	55211	99429	43169	66259	97786	59180
72	10127	46000	64984	75348	04115	33624	68774	60013	35515	62556
73	67995	81977	18984	64091	02785	27762	42529	97144	80407	64524
74	26304	80217	84934	82657	69291	35397	98714	35104	08187	48109
75	81994	41070	56642	64091	31229	02595	13513	45148	78722	30144
76	59537	34662	79631	89403	65212	09975	06118	86197	58208	16162
77	51228	10937	62396	81460	47331	91403	95007	06047	16846	64809
78	31089	37995	29577	07828	42272	54016	21950	86192	99046	84864
79	38207	97938	93459	75174	79460	55436	57206	87644	21296	43395
80	88666	31142	09474	89712	63153	62333	42212	06140	42594	43671
81	53365	56134	67582	92557	89520	33452	05134	70628	27612	33738
82	89807	74530	38004	90102	11693	90257	05500	79920	62700	43325
83	18682	81038	85662	90915	91631	22223	91588	80774	07716	12548
84	63571	32579	63942	25371	09234	94592	98475	76884	37635	33608
85	68927	56492	67799	95398	77642	54913	91853	08424	81450	76229
86	56401	63186	39389	88798	31356	89235	97036	32341	33292	73757
87	24333	95603	02359	72942	46287	95382	08452	62862	97869	71775
88	17025	84202	95199	62272	06366	16175	97577	99304	41587	03686
89	02804	08253	52133	20224	68034	50865	57868	22343	55111	03607
90	08298	03879	20995	19850	73090	13191	18963	82244	78479	99121
91	59883	01785	82403	96062	03785	03488	12970	64896	38336	30030
92	46982	06682	62864	91837	74021	89094	39952	64158	79614	78235
93	31121	47266	07661	02051	67599	24471	69843	83696	71402	76287
94	97867	56641	63416	17577	30161	87320	37752	73276	48969	41915
95	57364	86746	08415	14621	49430	22311	15836	72492	49372	44103
96	09559	26263	69511	28064	75999	44540	13337	10918	79846	54809
97	53873	55571	00608	42661	91332	63956	74087	59008	47493	99581
98	35531	19162	86406	05299	77511	24311	57257	22826	77555	05941
99	28229	88629	25695	94932	30721	16197	78742	34974	97528	45447

APPENDIX E

Critical Values of *t*

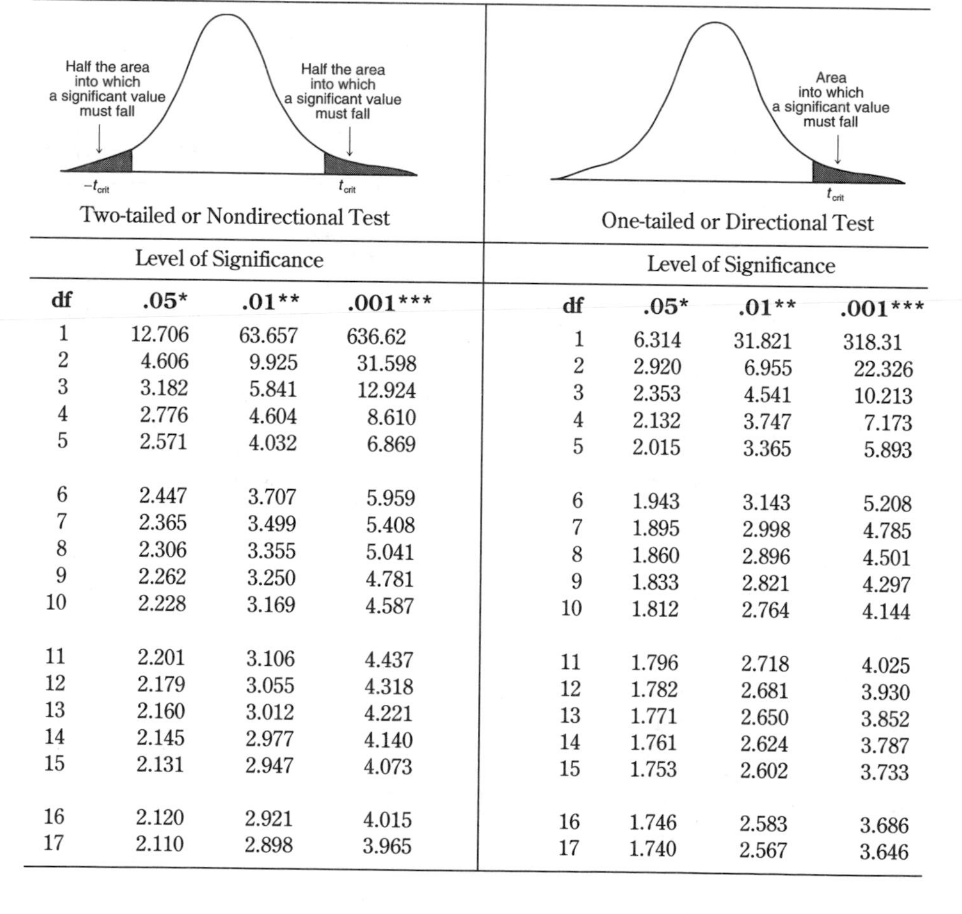

Two-tailed or Nondirectional Test

df	Level of Significance .05*	.01**	.001***
1	12.706	63.657	636.62
2	4.606	9.925	31.598
3	3.182	5.841	12.924
4	2.776	4.604	8.610
5	2.571	4.032	6.869
6	2.447	3.707	5.959
7	2.365	3.499	5.408
8	2.306	3.355	5.041
9	2.262	3.250	4.781
10	2.228	3.169	4.587
11	2.201	3.106	4.437
12	2.179	3.055	4.318
13	2.160	3.012	4.221
14	2.145	2.977	4.140
15	2.131	2.947	4.073
16	2.120	2.921	4.015
17	2.110	2.898	3.965

One-tailed or Directional Test

df	Level of Significance .05*	.01**	.001***
1	6.314	31.821	318.31
2	2.920	6.955	22.326
3	2.353	4.541	10.213
4	2.132	3.747	7.173
5	2.015	3.365	5.893
6	1.943	3.143	5.208
7	1.895	2.998	4.785
8	1.860	2.896	4.501
9	1.833	2.821	4.297
10	1.812	2.764	4.144
11	1.796	2.718	4.025
12	1.782	2.681	3.930
13	1.771	2.650	3.852
14	1.761	2.624	3.787
15	1.753	2.602	3.733
16	1.746	2.583	3.686
17	1.740	2.567	3.646

APPENDIX E (continued)

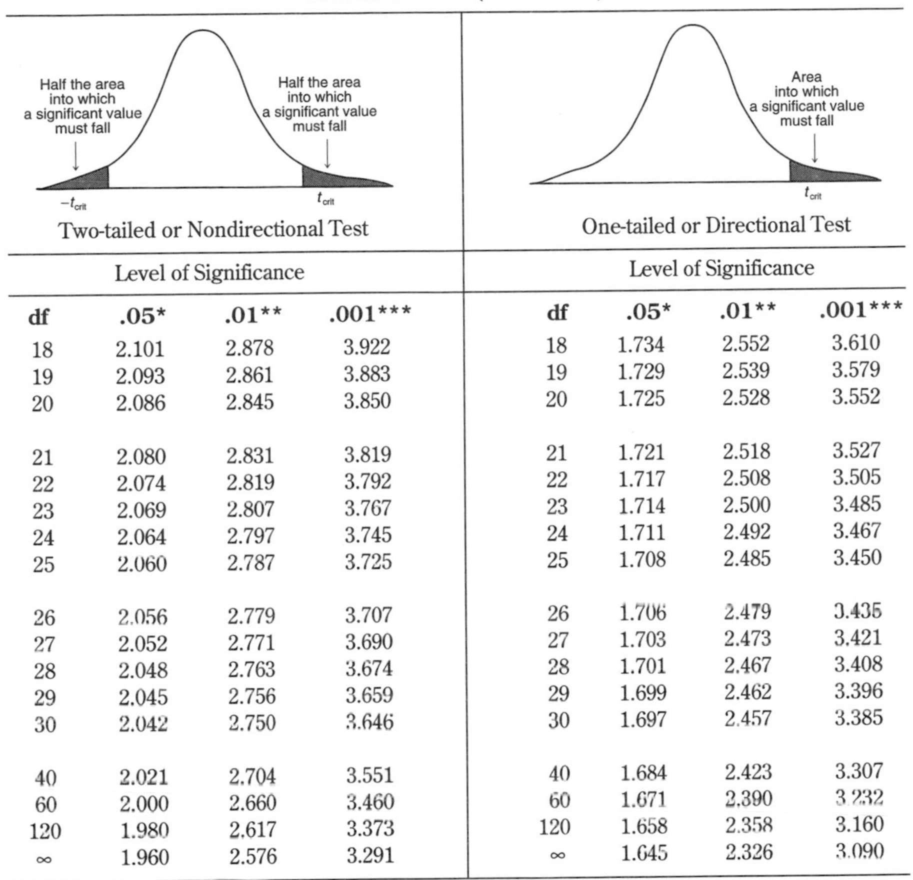

Two-tailed or Nondirectional Test

	Level of Significance		
df	**.05***	**.01****	**.001*****
18	2.101	2.878	3.922
19	2.093	2.861	3.883
20	2.086	2.845	3.850
21	2.080	2.831	3.819
22	2.074	2.819	3.792
23	2.069	2.807	3.767
24	2.064	2.797	3.745
25	2.060	2.787	3.725
26	2.056	2.779	3.707
27	2.052	2.771	3.690
28	2.048	2.763	3.674
29	2.045	2.756	3.659
30	2.042	2.750	3.646
40	2.021	2.704	3.551
60	2.000	2.660	3.460
120	1.980	2.617	3.373
∞	1.960	2.576	3.291

One-tailed or Directional Test

	Level of Significance		
df	**.05***	**.01****	**.001*****
18	1.734	2.552	3.610
19	1.729	2.539	3.579
20	1.725	2.528	3.552
21	1.721	2.518	3.527
22	1.717	2.508	3.505
23	1.714	2.500	3.485
24	1.711	2.492	3.467
25	1.708	2.485	3.450
26	1.706	2.479	3.435
27	1.703	2.473	3.421
28	1.701	2.467	3.408
29	1.699	2.462	3.396
30	1.697	2.457	3.385
40	1.684	2.423	3.307
60	1.671	2.390	3.232
120	1.658	2.358	3.160
∞	1.645	2.326	3.090

Using This Table

For a two-tailed test of significance (that is, when you have not predicted ahead of time which group will have higher scores or measurements than the other), use the values on the left side of the table. For a one-tailed test (when you knew beforehand which group ought to score higher), use the values on the right side.

If the two groups are independent of each other, df = $N_1 + N_2 - 2$; if the two groups are not independent (correlated), df = $N - 1$, where N is the number of pairs of scores.

APPENDIX F

Critical Values of F

Between group df

Within group df	1	2	3	4	5	6	7	8	9	10	11	12
1	161 **4,052**	200 **4,999**	216 **5,403**	225 **5,625**	230 **5,764**	234 **5,859**	237 **5,928**	239 **5,981**	241 **6,022**	242 **6,056**	243 **6,082**	244 **6,106**
2	18.51 **98.49**	19.00 **99.00**	19.16 **99.17**	19.25 **99.25**	19.30 **99.30**	19.33 **99.33**	19.36 **99.34**	19.37 **99.36**	19.38 **99.38**	19.39 **99.40**	19.40 **99.41**	19.41 **99.42**
3	10.13 **34.12**	9.55 **30.82**	9.28 **29.46**	9.12 **28.71**	9.01 **28.24**	8.94 **27.91**	8.88 **27.67**	8.84 **27.49**	8.81 **27.34**	8.78 **27.23**	8.76 **27.13**	8.74 **27.05**
4	7.71 **21.20**	6.94 **18.00**	6.59 **16.69**	6.39 **15.98**	6.26 **15.52**	6.16 **15.21**	6.09 **14.98**	6.04 **14.80**	6.00 **14.66**	5.96 **14.54**	5.93 **14.45**	5.91 **14.37**
5	6.61 **16.26**	5.79 **13.27**	5.41 **12.06**	5.19 **11.39**	5.05 **10.97**	4.95 **10.67**	4.88 **10.45**	4.82 **10.27**	4.78 **10.15**	4.74 **10.05**	4.70 **9.96**	4.68 **9.89**
6	5.99 **13.74**	5.14 **10.92**	4.76 **9.78**	4.53 **9.15**	4.39 **8.75**	4.28 **8.47**	4.21 **8.26**	4.15 **8.10**	4.10 **7.98**	4.06 **7.37**	4.03 **7.79**	4.00 **7.72**
7	5.59 **12.25**	4.74 **9.55**	4.35 **8.45**	4.12 **7.85**	3.97 **7.46**	3.87 **7.19**	3.79 **7.00**	3.73 **6.84**	3.68 **6.71**	3.63 **6.62**	3.60 **6.54**	3.57 **6.47**
8	5.32 **11.26**	4.46 **8.65**	4.07 **7.59**	3.84 **7.01**	3.69 **6.63**	3.58 **6.37**	3.50 **6.19**	3.44 **6.03**	3.39 **5.91**	3.34 **5.82**	3.31 **5.74**	3.28 **5.67**
9	5.12 **10.56**	4.26 **8.02**	3.86 **6.99**	3.63 **6.42**	3.48 **6.06**	3.37 **5.80**	3.29 **5.62**	3.23 **5.47**	3.18 **5.35**	3.13 **5.26**	3.10 **5.18**	3.07 **5.11**
10	4.96 **10.04**	4.10 **7.56**	3.71 **6.55**	3.48 **5.99**	3.33 **5.64**	3.22 **5.39**	3.14 **5.21**	3.07 **5.06**	3.02 **4.95**	2.97 **4.85**	2.94 **4.78**	2.91 **4.71**
11	4.84 **9.65**	3.98 **7.20**	3.59 **6.22**	3.36 **5.67**	3.20 **5.32**	3.09 **5.07**	3.01 **4.88**	2.95 **4.74**	2.90 **4.63**	2.86 **4.54**	2.82 **4.46**	2.79 **4.40**
12	4.75 **9.33**	3.88 **6.93**	3.49 **5.95**	3.26 **5.41**	3.11 **5.06**	3.00 **4.82**	2.92 **4.65**	2.85 **4.50**	2.80 **4.39**	2.76 **4.30**	2.72 **4.22**	2.69 **4.16**
13	4.67 **9.07**	3.80 **6.70**	3.41 **5.74**	3.18 **5.20**	3.02 **4.86**	2.92 **4.62**	2.84 **4.44**	2.77 **4.30**	2.72 **4.19**	2.67 **4.10**	2.63 **4.02**	2.60 **3.96**
14	4.60 **8.86**	3.74 **6.51**	3.34 **5.56**	3.11 **5.03**	2.96 **4.69**	2.85 **4.46**	2.77 **4.28**	2.70 **4.14**	2.65 **4.03**	2.60 **3.94**	2.56 **3.86**	2.53 **3.80**
15	4.54 **8.68**	3.68 **6.36**	3.29 **5.42**	3.06 **4.89**	2.90 **4.56**	2.79 **4.32**	2.70 **4.14**	2.64 **4.00**	2.59 **3.89**	2.55 **3.80**	2.51 **3.73**	2.48 **3.67**
16	4.49 **8.53**	3.63 **6.23**	3.24 **5.29**	3.01 **4.77**	2.85 **4.44**	2.74 **4.20**	2.66 **4.03**	2.59 **3.89**	2.54 **3.78**	2.49 **3.69**	2.45 **3.61**	2.42 **3.55**

Within group df	14	16	20	24	30	40	50	75	100	200	500	∞	
1	245 **6,142**	246 **6,169**	248 **6,208**	249 **6,234**	250 **6,258**	251 **6,286**	252 **6,302**	253 **6,323**	253 **6,334**	254 **6,352**	254 **6,361**	254 **6,366**	1
2	19.42 **99.43**	19.43 **99.44**	19.44 **99.45**	19.45 **99.46**	19.46 **99.47**	19.47 **99.48**	19.47 **99.48**	19.48 **99.49**	19.49 **99.49**	19.49 **99.49**	19.50 **99.50**	19.50 **99.50**	2
3	8.71 **26.92**	8.69 **26.83**	8.66 **26.69**	8.64 **26.60**	8.62 **26.50**	8.60 **26.41**	8.58 **26.35**	8.57 **26.27**	8.56 **26.23**	8.54 **26.18**	8.54 **26.14**	8.53 **26.12**	3
4	5.87 **14.24**	5.84 **14.15**	5.80 **14.02**	5.77 **13.93**	5.74 **13.83**	5.71 **13.74**	5.70 **13.69**	5.68 **13.61**	5.66 **13.57**	5.65 **13.52**	5.64 **13.48**	5.63 **13.46**	4
5	4.64 **9.77**	4.60 **9.68**	4.56 **9.55**	4.53 **9.47**	4.50 **9.38**	4.46 **9.29**	4.44 **9.24**	4.42 **9.17**	4.40 **9.13**	4.38 **9.07**	4.37 **9.04**	4.36 **9.02**	5
6	3.96 **7.60**	3.92 **7.52**	3.87 **7.39**	3.84 **7.31**	3.81 **7.23**	3.77 **7.14**	3.75 **7.09**	3.72 **7.02**	3.71 **6.99**	3.69 **6.94**	3.68 **6.90**	3.67 **6.88**	6
7	3.52 **6.35**	3.49 **6.27**	3.44 **6.15**	3.41 **6.07**	3.38 **5.98**	3.34 **5.90**	3.32 **5.85**	3.29 **5.78**	3.28 **5.75**	3.25 **5.70**	3.24 **5.67**	3.23 **5.65**	7
8	3.23 **5.56**	3.20 **5.48**	3.15 **5.36**	3.12 **5.28**	3.08 **5.20**	3.05 **5.11**	3.03 **5.06**	3.00 **5.00**	2.98 **4.96**	2.96 **4.91**	2.94 **4.88**	2.93 **4.86**	8
9	3.02 **5.00**	2.98 **4.92**	2.93 **4.80**	2.90 **4.73**	2.86 **4.64**	2.82 **4.56**	2.80 **4.51**	2.77 **4.45**	2.76 **4.41**	2.73 **4.36**	2.72 **4.33**	2.71 **4.31**	9
10	2.86 **4.60**	2.82 **4.52**	2.77 **4.41**	2.74 **4.33**	2.70 **4.25**	2.67 **4.17**	2.64 **4.12**	2.61 **4.05**	2.59 **4.01**	2.56 **3.96**	2.55 **3.93**	2.54 **3.91**	10
11	2.74 **4.29**	2.70 **4.21**	2.65 **4.10**	2.61 **4.02**	2.57 **3.94**	2.53 **3.86**	2.50 **3.80**	2.47 **3.74**	2.45 **3.70**	2.42 **3.66**	2.41 **3.62**	2.40 **3.60**	11
12	2.64 **4.05**	2.60 **3.98**	2.54 **3.86**	2.50 **3.78**	2.46 **3.70**	2.42 **3.61**	2.40 **3.56**	2.36 **3.49**	2.35 **3.46**	2.32 **3.41**	2.31 **3.38**	2.30 **3.36**	12
13	2.55 **3.85**	2.51 **3.78**	2.46 **3.67**	2.42 **3.59**	2.38 **3.51**	2.34 **3.42**	2.32 **3.37**	2.28 **3.30**	2.26 **3.27**	2.24 **3.21**	2.22 **3.18**	2.21 **3.16**	13
14	2.48 **3.70**	2.44 **3.62**	2.39 **3.51**	2.35 **3.43**	2.31 **3.34**	2.27 **3.26**	2.24 **3.21**	2.21 **3.14**	2.19 **3.11**	2.16 **3.06**	2.14 **3.02**	2.13 **3.00**	14
15	2.43 **3.56**	2.39 **3.48**	2.33 **3.36**	2.29 **3.29**	2.25 **3.20**	2.21 **3.12**	2.18 **3.07**	2.15 **3.00**	2.12 **2.97**	2.10 **2.92**	2.08 **2.89**	2.07 **2.87**	15
16	2.37 **3.45**	2.33 **3.37**	2.28 **3.25**	2.24 **3.18**	2.20 **3.10**	2.16 **3.01**	2.13 **2.96**	2.09 **2.89**	2.07 **2.86**	2.04 **2.80**	2.02 **2.77**	2.01 **2.75**	16

Adapted from *Statistical Methods* (6th ed.), by G. W. Snedecor and W. G. Cochran. .05 level (lightface type) and .01 level (**boldface type**).

Between group df (Continued)

Within group df	1	2	3	4	5	6	7	8	9	10	11	12	14	16	20	24	30	40	50	75	100	200	500	∞	
17	4.45 **8.40**	3.59 **6.11**	3.20 **5.18**	2.96 **4.67**	2.81 **4.34**	2.70 **4.10**	2.62 **3.93**	2.55 **3.79**	2.50 **3.68**	2.45 **3.59**	2.41 **3.52**	2.38 **3.45**	2.33 **3.35**	2.29 **3.27**	2.23 **3.16**	2.19 **3.08**	2.15 **3.00**	2.11 **2.92**	2.08 **2.86**	2.04 **2.79**	2.02 **2.76**	1.99 **2.70**	1.97 **2.67**	1.96 **2.65**	17
18	4.41 **8.28**	3.55 **6.01**	3.16 **5.09**	2.93 **4.58**	2.77 **4.25**	2.66 **4.01**	2.58 **3.85**	2.51 **3.71**	2.46 **3.60**	2.41 **3.51**	2.37 **3.44**	2.34 **3.37**	2.29 **3.27**	2.25 **3.19**	2.19 **3.07**	2.15 **3.00**	2.11 **2.91**	2.07 **2.83**	2.04 **2.78**	2.00 **2.71**	1.98 **2.68**	1.95 **2.62**	1.93 **2.59**	1.92 **2.57**	18
19	4.38 **8.18**	3.52 **5.93**	3.13 **5.01**	2.90 **4.50**	2.74 **4.17**	2.63 **3.94**	2.55 **3.77**	2.48 **3.63**	2.43 **3.52**	2.38 **3.43**	2.34 **3.36**	2.31 **3.30**	2.26 **3.19**	2.21 **3.12**	2.15 **3.00**	2.11 **2.92**	2.07 **2.84**	2.02 **2.76**	2.00 **2.70**	1.96 **2.63**	1.94 **2.60**	1.91 **2.54**	1.90 **2.51**	1.88 **2.49**	19
20	4.35 **8.10**	3.49 **5.85**	3.10 **4.94**	2.87 **4.43**	2.71 **4.10**	2.60 **3.87**	2.52 **3.71**	2.45 **3.56**	2.40 **3.45**	2.35 **3.37**	2.31 **3.30**	2.28 **3.23**	2.23 **3.13**	2.18 **3.05**	2.12 **2.94**	2.08 **2.86**	2.04 **2.77**	1.99 **2.69**	1.96 **2.63**	1.92 **2.56**	1.90 **2.53**	1.87 **2.47**	1.85 **2.44**	1.84 **2.42**	20
21	4.32 **8.02**	3.47 **5.78**	3.07 **4.87**	2.84 **4.37**	2.68 **4.04**	2.57 **3.81**	2.49 **3.65**	2.42 **3.51**	2.37 **3.40**	2.32 **3.31**	2.28 **3.24**	2.25 **3.17**	2.20 **3.07**	2.15 **2.99**	2.09 **2.88**	2.05 **2.80**	2.00 **2.72**	1.96 **2.63**	1.93 **2.58**	1.89 **2.51**	1.87 **2.47**	1.84 **2.42**	1.82 **2.38**	1.81 **2.36**	21
22	4.30 **7.94**	3.44 **5.72**	3.05 **4.82**	2.82 **4.31**	2.66 **3.99**	2.55 **3.76**	2.47 **3.59**	2.40 **3.45**	2.35 **3.35**	2.30 **3.26**	2.26 **3.18**	2.23 **3.12**	2.18 **3.02**	2.13 **2.94**	2.07 **2.83**	2.03 **2.75**	1.98 **2.67**	1.93 **2.58**	1.91 **2.53**	1.87 **2.46**	1.84 **2.42**	1.81 **2.37**	1.80 **2.33**	1.78 **2.31**	22
23	4.28 **7.88**	3.42 **5.66**	3.03 **4.76**	2.80 **4.26**	2.64 **3.94**	2.53 **3.71**	2.45 **3.54**	2.38 **3.41**	2.32 **3.30**	2.28 **3.21**	2.24 **3.14**	2.20 **3.07**	2.14 **2.97**	2.10 **2.89**	2.04 **2.78**	2.00 **2.70**	1.96 **2.62**	1.91 **2.53**	1.88 **2.48**	1.84 **2.41**	1.82 **2.37**	1.79 **2.32**	1.77 **2.28**	1.76 **2.26**	23
24	4.26 **7.82**	3.40 **5.61**	3.01 **4.72**	2.78 **4.22**	2.62 **3.90**	2.51 **3.67**	2.43 **3.50**	2.36 **3.36**	2.30 **3.25**	2.26 **3.17**	2.22 **3.09**	2.18 **3.03**	2.13 **2.93**	2.09 **2.85**	2.02 **2.74**	1.98 **2.66**	1.94 **2.58**	1.89 **2.49**	1.86 **2.44**	1.82 **2.36**	1.80 **2.33**	1.76 **2.27**	1.74 **2.23**	1.73 **2.21**	24
25	4.24 **7.77**	3.38 **5.57**	2.99 **4.68**	2.76 **4.18**	2.60 **3.86**	2.49 **3.63**	2.41 **3.46**	2.34 **3.32**	2.28 **3.21**	2.24 **3.13**	2.20 **3.05**	2.16 **2.99**	2.11 **2.89**	2.06 **2.81**	2.00 **2.70**	1.96 **2.62**	1.92 **2.54**	1.87 **2.45**	1.84 **2.40**	1.80 **2.32**	1.77 **2.29**	1.74 **2.23**	1.72 **2.19**	1.71 **2.17**	25
26	4.22 **7.72**	3.37 **5.53**	2.98 **4.64**	2.74 **4.14**	2.59 **3.82**	2.47 **3.59**	2.39 **3.42**	2.32 **3.29**	2.27 **3.17**	2.22 **3.09**	2.18 **3.02**	2.15 **2.96**	2.10 **2.86**	2.05 **2.77**	1.99 **2.66**	1.95 **2.58**	1.90 **2.50**	1.85 **2.41**	1.82 **2.36**	1.78 **2.28**	1.76 **2.25**	1.72 **2.19**	1.70 **2.15**	1.69 **2.13**	26
27	4.21 **7.68**	3.35 **5.49**	2.96 **4.60**	2.73 **4.11**	2.57 **3.79**	2.46 **3.56**	2.37 **3.39**	2.30 **3.26**	2.25 **3.14**	2.20 **3.06**	2.16 **2.98**	2.13 **2.93**	2.08 **2.83**	2.03 **2.74**	1.97 **2.63**	1.93 **2.55**	1.88 **2.47**	1.84 **2.38**	1.80 **2.33**	1.76 **2.25**	1.74 **2.21**	1.71 **2.16**	1.68 **2.12**	1.67 **2.10**	27
28	4.20 **7.64**	3.34 **5.45**	2.95 **4.57**	2.71 **4.07**	2.56 **3.76**	2.44 **3.53**	2.36 **3.36**	2.29 **3.23**	2.24 **3.11**	2.19 **3.03**	2.15 **2.95**	2.12 **2.90**	2.06 **2.80**	2.02 **2.71**	1.96 **2.60**	1.91 **2.52**	1.87 **2.44**	1.81 **2.35**	1.78 **2.30**	1.75 **2.22**	1.72 **2.18**	1.69 **2.13**	1.67 **2.09**	1.65 **2.06**	28
29	4.18 **7.60**	3.33 **5.42**	2.93 **4.54**	2.70 **4.04**	2.54 **3.73**	2.43 **3.50**	2.35 **3.33**	2.28 **3.20**	2.22 **3.08**	2.18 **3.00**	2.14 **2.92**	2.10 **2.87**	2.05 **2.77**	2.00 **2.68**	1.94 **2.57**	1.90 **2.49**	1.85 **2.41**	1.80 **2.32**	1.77 **2.27**	1.73 **2.19**	1.71 **2.15**	1.68 **2.10**	1.65 **2.06**	1.64 **2.03**	29
30	4.17 **7.56**	3.32 **5.39**	2.92 **4.51**	2.69 **4.02**	2.53 **3.70**	2.42 **3.47**	2.34 **3.30**	2.27 **3.17**	2.21 **3.06**	2.16 **2.98**	2.12 **2.90**	2.09 **2.84**	2.04 **2.74**	1.99 **2.66**	1.93 **2.55**	1.89 **2.47**	1.84 **2.38**	1.79 **2.29**	1.76 **2.24**	1.72 **2.16**	1.69 **2.13**	1.66 **2.07**	1.64 **2.03**	1.62 **2.01**	30
32	4.15 **7.50**	3.30 **5.34**	2.90 **4.46**	2.67 **3.97**	2.51 **3.66**	2.40 **3.42**	2.32 **3.25**	2.25 **3.12**	2.19 **3.01**	2.14 **2.94**	2.10 **2.86**	2.07 **2.80**	2.02 **2.70**	1.97 **2.62**	1.91 **2.51**	1.86 **2.42**	1.82 **2.34**	1.76 **2.25**	1.74 **2.20**	1.69 **2.12**	1.67 **2.08**	1.64 **2.02**	1.61 **1.98**	1.59 **1.96**	32
34	4.13 **7.44**	3.28 **5.29**	2.88 **4.42**	2.65 **3.93**	2.49 **3.61**	2.38 **3.38**	2.30 **3.21**	2.23 **3.08**	2.17 **2.97**	2.12 **2.89**	2.08 **2.82**	2.05 **2.76**	2.00 **2.66**	1.95 **2.58**	1.89 **2.47**	1.84 **2.38**	1.80 **2.30**	1.74 **2.21**	1.71 **2.15**	1.67 **2.08**	1.64 **2.04**	1.61 **1.98**	1.59 **1.94**	1.57 **1.91**	34
36	4.11 **7.39**	3.26 **5.25**	2.86 **4.38**	2.63 **3.89**	2.48 **3.58**	2.36 **3.35**	2.28 **3.18**	2.21 **3.04**	2.15 **2.94**	2.10 **2.86**	2.06 **2.78**	2.03 **2.72**	1.98 **2.62**	1.93 **2.54**	1.87 **2.43**	1.82 **2.35**	1.78 **2.26**	1.72 **2.17**	1.69 **2.12**	1.65 **2.04**	1.62 **2.00**	1.59 **1.94**	1.56 **1.90**	1.55 **1.87**	36
38	4.10 **7.35**	3.25 **5.21**	2.85 **4.34**	2.62 **3.86**	2.46 **3.54**	2.35 **3.32**	2.26 **3.15**	2.19 **3.02**	2.14 **2.91**	2.09 **2.82**	2.05 **2.75**	2.02 **2.69**	1.96 **2.59**	1.92 **2.51**	1.85 **2.40**	1.80 **2.32**	1.76 **2.22**	1.71 **2.14**	1.67 **2.08**	1.63 **2.00**	1.60 **1.97**	1.57 **1.90**	1.54 **1.86**	1.53 **1.84**	38
40	4.08 **7.31**	3.23 **5.18**	2.84 **4.31**	2.61 **3.83**	2.45 **3.51**	2.34 **3.29**	2.25 **3.12**	2.18 **2.99**	2.12 **2.88**	2.07 **2.80**	2.04 **2.73**	2.00 **2.66**	1.95 **2.56**	1.90 **2.49**	1.84 **2.37**	1.79 **2.29**	1.74 **2.20**	1.69 **2.11**	1.66 **2.05**	1.61 **1.97**	1.59 **1.94**	1.55 **1.88**	1.53 **1.84**	1.51 **1.81**	40

Between group df (Continued)

Within group df	1	2	3	4	5	6	7	8	9	10	11	12	14	16	20	24	30	40	50	75	100	200	500	∞	
42	4.07 **7.27**	3.22 **5.15**	2.83 **4.29**	2.59 **3.80**	2.44 **3.49**	2.32 **3.26**	2.24 **3.10**	2.17 **2.96**	2.11 **2.86**	2.06 **2.77**	2.02 **2.70**	1.99 **2.64**	1.94 **2.54**	1.89 **2.46**	1.82 **2.35**	1.78 **2.26**	1.73 **2.17**	1.68 **2.08**	1.64 **2.02**	1.60 **1.94**	1.57 **1.91**	1.54 **1.85**	1.51 **1.80**	1.49 **1.78**	42
44	4.06 **7.24**	3.21 **5.12**	2.82 **4.26**	2.58 **3.78**	2.43 **3.46**	2.31 **3.24**	2.23 **3.07**	2.16 **2.94**	2.10 **2.84**	2.05 **2.75**	2.01 **2.68**	1.98 **2.62**	1.92 **2.52**	1.88 **2.44**	1.81 **2.32**	1.76 **2.24**	1.72 **2.15**	1.66 **2.06**	1.63 **2.00**	1.58 **1.92**	1.56 **1.88**	1.52 **1.82**	1.50 **1.78**	1.48 **1.75**	44
46	4.05 **7.21**	3.20 **5.10**	2.81 **4.24**	2.57 **3.76**	2.42 **3.44**	2.30 **3.22**	2.22 **3.05**	2.14 **2.92**	2.09 **2.82**	2.04 **2.73**	2.00 **2.66**	1.97 **2.60**	1.91 **2.50**	1.87 **2.42**	1.80 **2.30**	1.75 **2.22**	1.71 **2.13**	1.65 **2.04**	1.62 **1.98**	1.57 **1.90**	1.54 **1.86**	1.51 **1.80**	1.48 **1.76**	1.46 **1.72**	46
48	4.04 **7.19**	3.19 **5.08**	2.80 **4.22**	2.56 **3.74**	2.41 **3.42**	2.30 **3.20**	2.21 **3.04**	2.14 **2.90**	2.08 **2.80**	2.03 **2.71**	1.99 **2.64**	1.96 **2.58**	1.90 **2.48**	1.86 **2.40**	1.79 **2.28**	1.74 **2.20**	1.70 **2.11**	1.64 **2.02**	1.61 **1.96**	1.56 **1.88**	1.53 **1.84**	1.50 **1.78**	1.47 **1.73**	1.45 **1.70**	48
50	4.03 **7.17**	3.18 **5.06**	2.79 **4.20**	2.56 **3.72**	2.40 **3.41**	2.29 **3.18**	2.20 **3.02**	2.13 **2.88**	2.07 **2.78**	2.02 **2.70**	1.98 **2.62**	1.95 **2.56**	1.90 **2.46**	1.85 **2.39**	1.78 **2.26**	1.74 **2.18**	1.69 **2.10**	1.63 **2.00**	1.60 **1.94**	1.55 **1.86**	1.52 **1.82**	1.48 **1.76**	1.46 **1.71**	1.44 **1.68**	50
55	4.02 **7.12**	3.17 **5.01**	2.78 **4.16**	2.54 **3.68**	2.38 **3.37**	2.27 **3.15**	2.18 **2.98**	2.11 **2.85**	2.05 **2.75**	2.00 **2.66**	1.97 **2.59**	1.93 **2.53**	1.88 **2.43**	1.83 **2.35**	1.76 **2.23**	1.72 **2.15**	1.67 **2.06**	1.61 **1.96**	1.58 **1.90**	1.52 **1.82**	1.50 **1.78**	1.46 **1.71**	1.43 **1.66**	1.41 **1.64**	55
60	4.00 **7.08**	3.15 **4.98**	2.76 **4.13**	2.52 **3.65**	2.37 **3.34**	2.25 **3.12**	2.17 **2.95**	2.10 **2.82**	2.04 **2.72**	1.99 **2.63**	1.95 **2.56**	1.92 **2.50**	1.86 **2.40**	1.81 **2.32**	1.75 **2.20**	1.70 **2.12**	1.65 **2.03**	1.59 **1.93**	1.56 **1.87**	1.50 **1.79**	1.48 **1.74**	1.44 **1.68**	1.41 **1.63**	1.39 **1.60**	60
65	3.99 **7.04**	3.14 **4.95**	2.75 **4.10**	2.51 **3.62**	2.36 **3.31**	2.24 **3.09**	2.15 **2.93**	2.08 **2.79**	2.02 **2.70**	1.98 **2.61**	1.94 **2.54**	1.90 **2.47**	1.85 **2.37**	1.80 **2.30**	1.73 **2.18**	1.68 **2.09**	1.63 **2.00**	1.57 **1.90**	1.54 **1.84**	1.49 **1.76**	1.46 **1.71**	1.42 **1.64**	1.39 **1.60**	1.37 **1.56**	65
70	3.98 **7.01**	3.13 **4.92**	2.74 **4.08**	2.50 **3.60**	2.35 **3.29**	2.23 **3.07**	2.14 **2.91**	2.07 **2.77**	2.01 **2.67**	1.97 **2.59**	1.93 **2.51**	1.89 **2.45**	1.84 **2.35**	1.79 **2.28**	1.72 **2.15**	1.67 **2.07**	1.62 **1.98**	1.56 **1.88**	1.53 **1.82**	1.47 **1.74**	1.45 **1.69**	1.40 **1.62**	1.37 **1.56**	1.35 **1.53**	70
80	3.96 **6.96**	3.11 **4.88**	2.72 **4.04**	2.48 **3.56**	2.33 **3.25**	2.21 **3.04**	2.12 **2.87**	2.05 **2.74**	1.99 **2.64**	1.95 **2.55**	1.91 **2.48**	1.88 **2.41**	1.82 **2.32**	1.77 **2.24**	1.70 **2.11**	1.65 **2.03**	1.60 **1.94**	1.54 **1.84**	1.51 **1.78**	1.45 **1.70**	1.42 **1.65**	1.38 **1.57**	1.35 **1.52**	1.32 **1.49**	80
100	3.94 **6.90**	3.09 **4.82**	2.70 **3.98**	2.46 **3.51**	2.30 **3.20**	2.19 **2.99**	2.10 **2.82**	2.03 **2.69**	1.97 **2.59**	1.92 **2.51**	1.88 **2.43**	1.85 **2.36**	1.79 **2.26**	1.75 **2.19**	1.68 **2.06**	1.63 **1.98**	1.57 **1.89**	1.51 **1.79**	1.48 **1.73**	1.42 **1.64**	1.39 **1.59**	1.34 **1.51**	1.30 **1.46**	1.28 **1.43**	100
125	3.92 **6.84**	3.07 **4.78**	2.68 **3.94**	2.44 **3.47**	2.29 **3.17**	2.17 **2.95**	2.08 **2.79**	2.01 **2.65**	1.95 **2.56**	1.90 **2.47**	1.86 **2.40**	1.83 **2.33**	1.77 **2.23**	1.72 **2.15**	1.65 **2.03**	1.60 **1.94**	1.55 **1.85**	1.49 **1.75**	1.45 **1.68**	1.39 **1.59**	1.36 **1.54**	1.31 **1.46**	1.27 **1.40**	1.25 **1.37**	125
150	3.91 **6.81**	3.06 **4.75**	2.67 **3.91**	2.43 **3.44**	2.27 **3.14**	2.16 **2.92**	2.07 **2.76**	2.00 **2.62**	1.94 **2.53**	1.89 **2.44**	1.85 **2.37**	1.82 **2.30**	1.76 **2.20**	1.71 **2.12**	1.64 **2.00**	1.59 **1.91**	1.54 **1.83**	1.47 **1.72**	1.44 **1.66**	1.37 **1.56**	1.34 **1.51**	1.29 **1.43**	1.25 **1.37**	1.22 **1.33**	150
200	3.89 **6.76**	3.04 **4.71**	2.65 **3.88**	2.41 **3.41**	2.26 **3.11**	2.14 **2.90**	2.05 **2.73**	1.98 **2.60**	1.92 **2.50**	1.87 **2.41**	1.83 **2.34**	1.80 **2.28**	1.74 **2.17**	1.69 **2.09**	1.62 **1.97**	1.57 **1.88**	1.52 **1.79**	1.45 **1.69**	1.42 **1.62**	1.35 **1.53**	1.32 **1.48**	1.26 **1.39**	1.22 **1.33**	1.19 **1.28**	200
400	3.86 **6.70**	3.02 **4.66**	2.62 **3.83**	2.39 **3.36**	2.23 **3.06**	2.12 **2.85**	2.03 **2.69**	1.96 **2.55**	1.90 **2.46**	1.85 **2.37**	1.81 **2.29**	1.78 **2.23**	1.72 **2.12**	1.67 **2.04**	1.60 **1.92**	1.54 **1.84**	1.49 **1.74**	1.42 **1.64**	1.38 **1.57**	1.32 **1.47**	1.28 **1.42**	1.22 **1.32**	1.16 **1.24**	1.13 **1.19**	400
1000	3.85 **6.66**	3.00 **4.62**	2.61 **3.80**	2.38 **3.34**	2.22 **3.04**	2.10 **2.82**	2.02 **2.66**	1.95 **2.53**	1.89 **2.43**	1.84 **2.34**	1.80 **2.26**	1.76 **2.20**	1.70 **2.09**	1.65 **2.01**	1.58 **1.89**	1.53 **1.81**	1.47 **1.71**	1.41 **1.61**	1.36 **1.54**	1.30 **1.44**	1.26 **1.38**	1.19 **1.28**	1.13 **1.19**	1.08 **1.11**	1000
∞	3.84 **6.64**	2.99 **4.60**	2.60 **3.78**	2.37 **3.32**	2.21 **3.02**	2.09 **2.80**	2.01 **2.64**	1.94 **2.51**	1.88 **2.41**	1.83 **2.32**	1.79 **2.24**	1.75 **2.18**	1.69 **2.07**	1.64 **1.99**	1.57 **1.87**	1.52 **1.79**	1.46 **1.69**	1.40 **1.59**	1.35 **1.52**	1.28 **1.41**	1.24 **1.36**	1.17 **1.25**	1.11 **1.15**	1.00 **1.00**	∞

APPENDIX G

Critical Values of Chi Square (χ^2)

df	.01	.05	.10
1	6.64	3.84	2.71
2	9.21	5.99	4.60
3	11.34	7.82	6.25
4	13.28	9.49	7.78
5	15.09	11.07	9.24
6	16.81	12.59	10.64
7	18.48	14.07	12.02
8	20.09	15.51	13.36
9	21.67	16.92	14.68
10	23.21	18.31	15.99
11	24.72	19.68	17.28
12	26.22	21.03	18.55
13	27.69	22.36	19.81
14	29.14	23.68	21.06
15	30.58	25.00	22.31
16	32.00	26.97	23.54
17	33.41	27.59	24.77

Appendix G is abridged from Table IV of Fisher & Yates, *Statistical Tables for Biological, Agricultural, and Medical Research,* published by Oliver & Boyd Ltd., Edinburgh, and by permission of the authors and publishers.

(The significance level for each value is given at the top of the column.)

APPENDIX G (continued)

df	.01	.05	.10
18	34.80	28.87	25.99
19	36.19	30.14	27.20
20	37.57	31.41	28.41
21	38.93	32.67	29.62
22	40.29	33.92	30.81
23	41.64	35.17	32.01
24	42.98	36.42	33.20
25	44.31	37.65	34.38
26	45.64	38.88	35.56
27	46.96	40.11	36.74
28	48.28	41.34	37.92
29	49.59	42.56	39.09
30	50.89	43.77	40.26

APPENDIX H

Probabilities Associated with the Mann–Whitney *U* Test (One-tailed Test)

$N_2 = 3$

U \ N_1	1	2	3
0	.250	.100	.050
1	.500	.200	.100
2	.750	.400	.200
3		.600	.350
4			.500
5			.650

$N_2 = 4$

U \ N_1	1	2	3	4
0	.200	.067	.028	.014
1	.400	.133	.057	.029
2	.600	.267	.114	.057
3		.400	.200	.100
4		.600	.314	.171
5			.429	.243
6			.571	.343
7				.443
8				.557

APPENDIX H (continued)

$N_2 = 5$

U \ N_1	1	2	3	4	5
0	.167	.047	.018	.008	.004
1	.333	.095	.036	.016	.008
2	.500	.190	.071	.032	.016
3	.667	.286	.125	.056	.028
4		.429	.196	.095	.048
5		.571	.286	.143	.075
6			.393	.206	.111
7			.500	.278	.155
8			.607	.365	.210
9				.452	.274
10				.548	.345
11					.421
12					.500
13					.579

$N_2 = 6$

U \ N_1	1	2	3	4	5	6
0	.143	.036	.012	.005	.002	.001
1	.286	.071	.024	.010	.004	.002
2	.428	.143	.048	.019	.009	.004
3	.571	.214	.083	.033	.015	.008
4		.321	.131	.057	.026	.013
5		.429	.190	.086	.041	.021
6		.571	.274	.129	.063	.032
7			.357	.176	.089	.047
8			.452	.238	.123	.066
9			.548	.305	.165	.090
10				.381	.214	.120
11				.457	.268	.155
12				.545	.331	.197
13					.396	.242
14					.465	.294
15					.535	.350
16						.409
17						.469
18						.531

APPENDIX H (continued)

$N_2 = 7$

U \ N_1	1	2	3	4	5	6	7
0	.125	.028	.008	.003	.001	.001	.000
1	.250	.056	.017	.006	.003	.001	.001
2	.375	.111	.033	.012	.005	.002	.001
3	.500	.167	.058	.021	.009	.004	.002
4	.625	.250	.092	.036	.015	.007	.003
5		.333	.133	.055	.024	.011	.006
6		.444	.192	.082	.037	.017	.009
7		.556	.258	.115	.053	.026	.013
8			.333	.158	.074	.037	.019
9			.417	.206	.101	.051	.027
10			.500	.264	.134	.069	.036
11			.583	.324	.172	.090	.049
12				.394	.216	.117	.064
13				.464	.265	.147	.082
14				.538	.319	.183	.104
15					.378	.223	.130
16					.438	.267	.159
17					.500	.314	.191
18					.562	.365	.228
19						.418	.267
20						.473	.310
21						.527	.355
22							.402
23							.451
24							.500
25							.549

APPENDIX H (continued)

$N_2 = 8$

U \ N_1	1	2	3	4	5	6	7	8	t	Normal
0	.111	.022	.006	.002	.001	.000	.000	.000	3.308	.001
1	.222	.044	.012	.004	.002	.001	.000	.000	3.203	.001
2	.333	.089	.024	.008	.003	.001	.001	.000	3.098	.001
3	.444	.133	.042	.014	.005	.002	.001	.001	2.993	.001
4	.556	.200	.067	.024	.009	.004	.002	.001	2.888	.002
5		.267	.097	.036	.015	.006	.003	.001	2.783	.003
6		.356	.139	.055	.023	.010	.005	.002	2.678	.004
7		.444	.188	.077	.033	.015	.007	.003	2.573	.005
8		.556	.248	.107	.047	.021	.010	.005	2.468	.007
9			.315	.141	.064	.030	.014	.007	2.363	.009
10			.387	.184	.085	.041	.020	.010	2.258	.012
11			.461	.230	.111	.054	.027	.014	2.153	.016
12			.539	.285	.142	.071	.036	.019	2.048	.020
13				.341	.177	.091	.047	.025	1.943	.026
14				.404	.217	.114	.060	.032	1.838	.033
15				.467	.262	.141	.076	.041	1.733	.041
16				.533	.311	.172	.095	.052	1.628	.052
17					.362	.207	.116	.065	1.523	.064
18					.416	.245	.140	.080	1.418	.078
19					.472	.286	.168	.097	1.313	.094
20					.528	.331	.198	.117	1.208	.113
21						.377	.232	.139	1.102	.135
22						.426	.268	.164	.998	.159
23						.475	.306	.191	.893	.185
24						.525	.347	.221	.788	.215
25							.389	.253	.683	.247
26							.433	.287	.578	.282
27							.478	.323	.473	.318
28							.522	.360	.368	.356
29								.399	.263	.396
30								.439	.158	.437
31								.480	.052	.481
32								.520		

APPENDIX I

Critical Values of Spearman's Ranked Correlation Coefficient (r_S)

n	$\alpha = .05$	$\alpha = .025$	$\alpha = .01$	$\alpha = .005$
5	0.900	—	—	—
6	0.829	0.886	0.943	—
7	0.714	0.786	0.893	—
8	0.643	0.738	0.833	0.881
9	0.600	0.683	0.783	0.833
10	0.564	0.648	0.745	0.794
11	0.523	0.623	0.736	0.818
12	0.497	0.591	0.703	0.780
13	0.475	0.566	0.673	0.745
14	0.457	0.545	0.646	0.716
15	0.441	0.525	0.623	0.689
16	0.425	0.507	0.601	0.666
17	0.412	0.490	0.582	0.645
18	0.399	0.476	0.564	0.625
19	0.388	0.462	0.549	0.608
20	0.377	0.450	0.534	0.591
21	0.368	0.438	0.521	0.576
22	0.359	0.428	0.508	0.562
23	0.351	0.418	0.496	0.549
24	0.343	0.409	0.485	0.537
25	0.336	0.400	0.475	0.526
26	0.329	0.392	0.465	0.515
27	0.323	0.385	0.456	0.505
28	0.317	0.377	0.448	0.496
29	0.311	0.370	0.440	0.487
30	0.305	0.364	0.432	0.478

From "Distribution of Sums of Squares of Rank Differences for Small Numbers of Individuals," E. G. Olds, *Annals of Mathematical Statistics,* Volume 9 (1938). Reproduced with permission.

APPENDIX J

Overcoming Math Anxiety

If you are what might be termed a "math anxious" or "math avoidant" person, this section may be helpful to you. Most of the material in this chapter is drawn from the theory and practice of rational-emotive therapy (RET), originally developed by the psychologist Albert Ellis. RET has been shown through research to be quite effective in helping people overcome problems like yours. Unfortunately, in a book devoted to statistics, I can only introduce you to some of the basic ideas and techniques. If you are interested, you can enrich your understanding by reading books like Ellis and Harper's *A Guide to Rational Living* or Kranzler's *You Can Change How You Feel* (notice the sneaky way of getting a plug in?)

Fear of math, or math anxiety, is what is call a *debilitative* emotion. Debilitative emotions such as math anxiety are problem emotions because (1) they are extremely unpleasant, and (2) they tend to lead to self-defeating behavior, such as "freezing" on a test or avoiding courses or occupations you otherwise would enjoy.

What you do about your math anxiety (or any other problem) will depend on your theory of what is causing the problem. For example, some people believe that the cause is hereditary: "I get my fear of math from mother, who always had the same problem." Others believe that the cause lies in the environment: "Women are taught from a very young age that they are not supposed to be good in math, to avoid it, and to be afraid of it." The implication of these theories is that if the cause is hereditary you can't do much about the problem (you can't change your genetic structure), or if the cause is the culture in which you live, by the time you can change what society does to its young, it will still be too late to help you. While there may be some truth in both the heredity and environmental theories, I believe that they can, at most, set only general limits to your performance. Within these limits your performance can fluctuate considerably. Though you have very little power to change society and no ability to change your heredity, you still have enormous power to change yourself if you choose to do so, if you know how to bring about that change, and if you work hard at it.

Let's begin with the ABC's. *A* stands for *A*ctivating event or experience, such as taking a difficult math test; *C* stands for the emotional *C*onsequence, such as extreme nervousness. Most people seem to believe that A causes C. In fact, this theory seems to be built right into our language. Consider:

Activating Event	Causes	Emotional Consequence
(Something happens . . .	that causes me . . .	feelings)
"When you talk about math . . .	you make me . . .	so upset"
"This test . . .	makes me . . .	nervous"

The implications of this A-causes-C theory are (1) you can't help how you feel, and (2) the way to deal with the problem is to avoid or escape from activating events such as math tests.

But is the A-causes-C theory true? Respond to the following items by indicating how you would feel if you were to experience the event. Use a scale that ranges from –5, indicating extremely unpleasant emotions (such as rage, depression, or extreme anxiety), to +5, to indicate an emotion that is extremely positive (such as elation or ecstasy); or use a 0 if you would experience neutral, neither positive nor negative, feelings:

1. Handling snakes.
2. Giving a speech in front of one of your classes.
3. Seeing your eight-year-old son playing with dolls.
4. The death of a loved one in an automobile accident.

I have administered items like this to hundreds of people and have found that for items 1 through 3 the responses have ranged all the way from –5 to +5. On the item concerning the death of a loved one, most people respond with a –5, but when questioned, they have heard of cultures where even death is considered to be a positive event (in the United States everyone wants to go to heaven but nobody wants to die). Why is it that, given the same *A*ctivating event, people's emotional *C*onsequences vary so much?

Differing responses like this suggest that maybe $A \rightarrow C$ isn't the whole story. There must be something else, something that accounts for the different ways people respond to the same stimulus. I believe that it is not *A*, the activating event, that causes *C*, the emotional consequence. Rather, it is *B*, your Belief about *A*, that causes you to feel as you do at point *C*. Take the example of observing your eight-year-old son playing with dolls. What does the person who experiences feelings of joy believe about what he or she sees? Perhaps something like, "Isn't that wonderful! He's learning nurturing attitudes and tenderness. I really like that!" But the person who experiences very negative feelings probably is thinking, "Isn't that awful! If he keeps that up, he'll surely turn into an effeminate man, or even be gay, and that really would be terrible!"

Ellis has identified some specific beliefs that most of us have learned and that cause us a great deal of difficulty. He calls these beliefs "irrational beliefs." A number of these beliefs have particular relevance to the phenomenon of math anxiety:

> I must be competent and adequate in all possible respects if I am to consider myself to be a worthwhile person. (If I'm not good at math, I'm not a very smart person.)
>
> It's catastrophic when things are not the way I'd like them to be. (It's terrible and awful to have trouble with statistics.)
>
> When something seems dangerous or about to go wrong, I must constantly worry about it. (I can't control my worrying and fretting about statistics.)
>
> My unhappiness is externally caused. I can't help feeling and acting as I do and I can't change my feelings or actions. (Trying to do math simply makes me feel awful; that's just what it does to me.)
>
> Given my childhood experiences and the past I have had, I can't help being as I am today and I'll remain this way indefinitely. (I'll never change; that's just how I am.)
>
> I can't settle for less than the right or perfect solution to my problems. (Since I can't be a math whiz, there's no sense in trying to do math at all.)
>
> It is better for me to avoid than to face life's frustrations and difficulties. (Since math always makes me feel bad, the only sensible thing to do is to avoid math.)

Do any of these sound familiar? If they do, chances are good that you not only learned to believe them a long time ago, but also that you keep the belief going by means of self-talk. The first step in changing is to increase your awareness of the kind of self-talk that you do. When you think, you think with words, sentences, and images. If you pay attention to these cognitive events, you may notice one or more of the following types of self-talk, which may indicate your underlying irrational beliefs:

Catastrophizing

This type of self-talk is characterized by the use of terms or phrases such as "It's awful!" "It's terrible!" or "I can't stand it!" Now, there are some events that most of us would agree are extremely bad, such as bombing innocent people and earthquakes that kill thousands. Chances are good that you will never be the victim of such an event. But your mind is powerful: if you *believe* that your misfortunes are catastrophes, then you will *feel* accordingly. Telling yourself how catastrophic it is to do badly on a stats test will almost guarantee that you will feel awful about it. And that emotional response, in turn, can affect how you deal with the situation. It is appropriate to be *concerned* about doing well on a test, because concern motivates you to prepare for and to do your best. But when you are *overconcerned* you can make yourself so nervous that your performance goes down instead of up.

Do you see how all this relates to the first irrational belief on our list? Performing poorly on a stats test would be *awful* because you believe that you *must* be competent in all possible respects. If you were to fail at something important to you, that would make you a *failure:* someone who couldn't respect himself or herself. One of the oddest things about irrational beliefs like this is the uneven way we apply them. Your friend could bomb a test, and you'd still think him or her a worthwhile person. But do badly yourself, and the sky falls in!

When you indoctrinate yourself with catastrophic ideas, when you tell yourself over and over again how *horrible* it would be if you were to perform poorly, then you defeat yourself, because you become so anxious that you help bring about the very thing you're afraid of, or you avoid the experience that could benefit you.

Overgeneralizing Self-talk

When you overgeneralize, you take a bit of evidence and draw conclusions that go beyond the data. If you experienced difficulty with math as a child, you may have concluded, "I'll *never* be good at math" or "I'm stupid in math." If you failed a math course, then you tended to think of yourself as a failure who will never be able to succeed, and trying harder would be completely useless.

Rationally, though, failing once doesn't make you a "failure." Because you had difficulty in the past doesn't prove that you will never succeed. If it did, nobody would ever learn to walk!

The most pernicious form of overgeneralizing is self-evaluation. We have a tendency to tie up our feelings of self-worth with our performance. When we do well at something, we say, "Hey! I'm a pretty good (or competent or worthwhile) person!" But when we perform poorly, we tend to believe that we are now worthless as a person. This process begins in childhood. When Johnny does something we consider bad, we tend to encourage overgeneralization by saying, "Johnny, you are a *bad boy*" (that is, you are worth less as a person)!

If you were a worthless or stupid person, you wouldn't have gotten far enough in your education to be reading this book. True, in the past you may have had difficulty in math, and math may be difficult for you now. But how does that prove that you can't learn it? There is absolutely no evidence that your situation is hopeless or that it is useless to try. The only way to make it hopeless is to tell yourself, over and over, how hopeless it is.

Demanding Self-talk

This type of self-talk includes the use of words such as "should," "must," and "need." If you are math anxious, chances are that you use these words to beat up on yourself. You make a mistake and say, "I shouldn't have made that mistake! How could I have been so stupid?" I have a racquetball partner who informed me that he finds it difficult to concentrate on his work for the rest of the day after he has played poorly. He believes that he *should* have done better. Instead of being calmly regretful for having made some errors and thinking about how to do better next time, he bashes himself over the head psychologically for not doing perfectly well, every time.

"But," you may say, "I *need* to be successful," or "I *have to* pass this course." Have to? The first time? Or you can't survive? It would be nice to be successful because of the advantages it would bring you, but lots of people do manage to function in life even after doing badly on a statistics test. To the degree that you believe you *need* a certain level of performance, to that degree you will experience anxiety about possible failure and thereby increase the chance of failure.

HOW TO DEAL WITH MATH ANXIETY

What can you do about a way of thinking that seems so automatic, so ingrained? Here are a series of steps that will probably help. I'd suggest that you try them out, in order, even though you may not expect them to work for you. You might just be surprised!

Step 1. Record your feelings. When you notice that you are feeling anxious, guilty, angry, or depressed about some aspect of your statistics course, record your emotional experience. Describe your feelings as accurately as you can. You might write things such as, "I feel guilty about not having taken more math as an underclassman," or "I feel really nervous about the test we're having next week," or "I'm too shy to ask questions in class." Write down all the unpleasant feelings you have at the time. When you have done this, you will have described C, the emotional Consequence part of the ABC paradigm.

Step 2. Describe the Activating event or experience (A). Briefly write down what it was that seemed to trigger your feelings. Here are some common activating events for math anxiety. When you write your own, record the thing that is most likely to have happened.

> I was assigned some difficult statistics problems, and I don't know how to do them.
>
> I have a test coming up that I will almost surely fail.

I need more information about some of the material, but I'm afraid to ask about it in class because I'll look stupid.

Step 3. Identify your irrational beliefs. As accurately as you can, record what you were saying to yourself before and during the time when you experienced the emotions you recorded in step 1. The first few times you do this you may have difficulty, because you don't usually pay much attention to the thoughts that seem to race through your head. Although it is difficult to become aware of your thoughts, it is not impossible. One technique you can use is to ask yourself, "What must I have been saying to myself about A (the activating event) at point B in order to experience C (the emotional consequence)?

Suppose your first three steps looked like this:

Step 1. (Describing C, the emotional Consequence) I feel really nervous and miserable.

Step 2. (the Activating event, A) My advisor told me I need to take a statistics class.

Step 3. Identify B, the Belief that leads from A to C. Obviously, you're not saying, "Wow, I'm really going to enjoy that class!" You must have been saying something like:

"If I fail, that'll be *awful*!"
"I'll be a real failure!"
"I'll *never* be any good at statistics!
"What will the other students and the professor think of me if I do badly?"

Step 4. Challenge each of the beliefs you have identified. After you have written down your self-talk in step 3, look at each statement and dispute it. One question you can ask to test the rationality of any belief is, "Where's the evidence for this belief?" Let's look at each of the examples listed in step 3:

1. Where's the evidence that it will be awful if I fail? True, failure would be unfortunate, but would it be catastrophic? I'd better remember that if I'm overconcerned with doing well I will be even more likely to fail.
2. Where's the evidence that if I fail the test I, as a person, will be a failure? The worst I can possibly be is an FHB (a fallible human being) along with the rest of the human race.
3. Where's the evidence that I'll never be good in statistics? I may have some evidence that similar material was difficult for me in the past, but how can that prove anything about the future?

4. This statement appears to be a rhetorical question. Chances are that I'm not really wondering what others will think of me if I fail, but rather telling myself all the bad things they'll think—and how awful that will be. Both parts of this can be challenged: where's the evidence that they'll think bad things about me and, even if they do, would that be catastrophic?

Step 5. Once you have identified an irrational belief, the next step is to replace it with a rational one. Ask yourself what you would *rather* believe—what your best friend might believe—what Kathryn Hepburn or Jack Kennedy or Mahatma Gandhi probably would believe. Then, every time you find yourself moving into that old irrational self-talk, answer it with the new alternative belief.

Step 6. Do rational-emotive imagery. After you have practiced replacing your irrational beliefs a few times, you may feel better. Some people, however, report that they now understand that their beliefs cause their unpleasant emotions, and they realize that those beliefs are irrational, but they still feel much the same way as before. If this is true of you, you may benefit from doing some imagery. I will discuss both mastery and coping imagery techniques because some of my students have reported that one approach is more effective for them than the other. Before attempting either kind of imagery, however, do spend several days practicing steps 1 through 5.

Mastery Imagery. In the mastery imagery approach, you are to imagine yourself mastering the situation, that is, feeling and acting in an appropriate way in the presence of the activating event. If you are anxious about a statistics test, imagine feeling calm or at most slightly concerned while taking the test, answering the questions as well as you can, calmly leaving the exam, and being able to look back on the experience with some satisfaction. Imagine speaking rationally to yourself during the whole experience (taken from your material in step 5). If you feel very upset during the experience, terminate the imagery; go back and reread step 4 and attempt the imagery again the next day. For either kind of imagery to be effective, you will need to work at it for at least a half-hour per day for a week; don't expect immediate results.

Coping Imagery. Again imagine yourself in the experience that you're having problems with, for example, taking a statistics test. This time include having difficulty and starting to experience anxiety. Then imagine dealing with the anxiety by saying to yourself, "Stop! Relax!" Try to force yourself to feel more calm. Breathe deeply a few times, remind yourself of rational self-talk, and try to change the extremely anxious feelings to ones that are more calm. Imagine coping with the problem. Again, you won't experience immediate success; it usually takes at least a week of imagery work before you begin to get results.

Step 7. Activity homework. You can only live in your imagination so long if you want to attain objectives in the real world. Sooner or later you need to take a deep breath and DO SOMETHING. As you begin to make some conscious, real-world changes, be aware of your self-talk. When you notice yourself feeling emotionally upset, dispute your irrational beliefs as actively as you can. If things don't get better immediately, don't give up—keep using these techniques for at least a couple of weeks. Remember, the odds are in your favor!

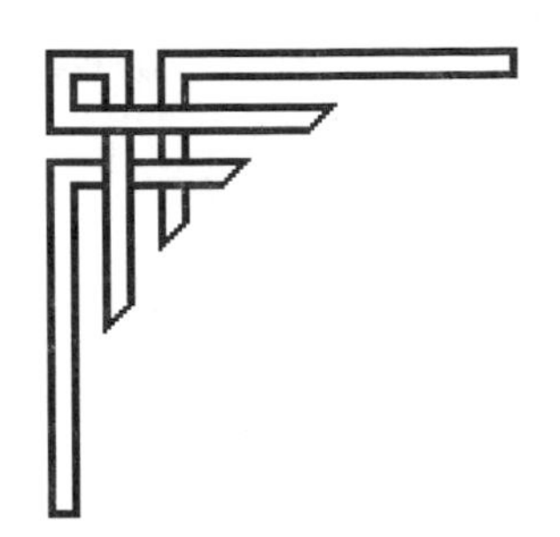

Index

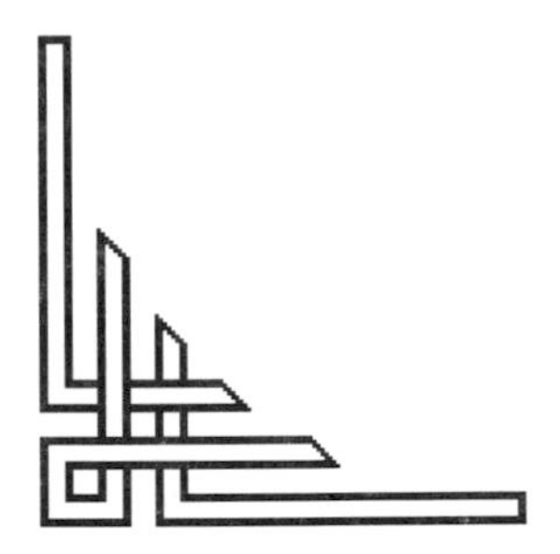